T0330205

Governance for Urban Sustainability and Resilience

Governance for Urban Sustainability and Resilience

Responding to Climate Change and the Relevance of the Built Environment

Jeroen van der Heijden

Assistant Professor, Australian National University, Regulatory Institutions Network (RegNet), Australia; and University of Amsterdam, Amsterdam Law School, the Netherlands

Edward Elgar

Cheltenham, UK • Northampton, MA, USA

Published by
Edward Elgar Publishing Limited
The Lypiatts
15 Lansdown Road
Cheltenham
Glos GL50 2JA
UK

Edward Elgar Publishing, Inc.
William Pratt House
9 Dewey Court
Northampton
Massachusetts 01060
USA

A catalogue record for this book
is available from the British Library

Library of Congress Control Number: 2014941546

This book is available electronically in the ElgarOnline.com Social and Political Sciences Subject Collection, E-ISBN 978 1 78254 813 3

ISBN 978 1 78254 812 6

Typeset by Columns Design XML Ltd, Reading
Printed and bound in Great Britain by T.J. International Ltd, Padstow

Contents

Biographical information

Dr Jeroen van der Heijden (1977) is Assistant Professor of Environmental Governance. He received an MSc in Architecture (high distinction equivalent) in 2002 and a PhD in Public Policy (highest honours) in 2009, both from the Delft University of Technology, the Netherlands. He currently holds positions at the Australian National University, Regulatory Institutions Network (RegNet) and the University of Amsterdam, Amsterdam Law School.

He has spent time teaching and researching at the above mentioned universities as well as at the University of Washington (United States), the University of Oxford (United Kingdom), the Sri Venkateswara University (India), the London School of Economics (United Kingdom), the University of East Anglia (United Kingdom) and the Queensland University of Technology (Australia).

Dr van der Heijden's main research interest is in the theory and practice of environmental governance. His current research is driven by questions such as: How can we achieve sustainability without the force of law? Why are free markets often unable to achieve a meaningful improvement of environmental performance? Is collaboration between governments, businesses and civil society a fruitful approach to address environmental problems?

In particular, he seeks to understand why the often normative assumptions of new, collaborative and experimental forms of environmental governance are not realized in practice. His work is international comparative in nature, and often builds on medium-n case-studies including large numbers of interviewees.

Dr van der Heijden has a track record of outstanding publications. Since 2008 he has published over 25 articles in peer-reviewed journals, and another ten in policy and practitioners' journals. He has published four books, a series of book chapters, and a wide range of publications for a policy and practitioner audience. He maintains an urban sustainability and resilience blog on which he regularly discusses his research findings (urbansustainabilityresilience.wordpress.com).

Before pursuing his PhD he worked for a number of years at a building consultancy firm. Here he advised city governments on the implementation of building regulation, and supported property developers and property owners in building projects. In 2012 he was awarded a prestigious VENI early career researcher's grant from the Dutch Organisation for Scientific Research (€250 000 – awarded to the top 10 per cent of grant applicants); the grant has allowed him, among others, to write the current book.

Dr van der Heijden is a citizen of the Netherlands and currently lives in Australia. He enjoys exploring the Australian bush, photographing Australia's big skies and gum-trees during the golden hours, and being overwhelmed time and again by the vast openness of this astonishing continent.

Preface

While writing this book, often in evenings and weekends when my office building was at its quietest, it often seemed as though the whole building was heated and lit just for me. I wondered: Why don't we have a small building where the lights and heat are on for those who prefer to work after 8 p.m., and on Saturdays and Sundays, and lock off the rest of the buildings? Rapidly a more important question developed: How much energy is being wasted?

I soon found the answer. In Australian office buildings 55 per cent of energy is consumed after office hours, during weekends and holidays (Greensense, 2013). In other words, most energy is consumed when hardly anyone uses these buildings.

The Australian National University has an internationally acclaimed, award winning governance programme in place, ANU-Green, that seeks to inform university staff and building managers on how to reduce their energy consumption. Still, the Australian National University has not implemented a rigorous set of governance tools to fully utilize the opportunities to increase environmental and resource sustainability on their campus and to capitalize on the related cost savings.

As much as I considered the energy consumption of my university, I didn't worry that my office building would suddenly collapse. The building is not that old yet, and in Canberra one does not normally experience earthquakes or hurricanes. Yet when considering building design and the environment this is a surprisingly important issue.

'Earthquakes don't kill people, poorly constructed buildings do!', Iwan Gunawan (2009) wrote in *The Jakarta Post* shortly after the devastating Mw 7.4 earthquake that hit Sumatra, Indonesia, on 30 September 2009. The earthquake killed more than 1100 people, injured close to 3000 and severely damaged or destroyed 135 000 houses. The same adagio holds for the bushfires (and floods) that seem to have been increasing in frequency and intensity in Australia over the last years (Cork, 2010).

I wondered, year after year, during bushfire season in Australia, why people still build houses made of wood and other fire-prone materials in areas vulnerable to bushfires. Although I swapped practising architecture for studying environmental governance a long time ago, the architect in me very well recalls that the materials and designs are in place to make buildings highly fire resistant. This raises the question: Why don't government or insurance companies mandate the use of these in Australia? This would result in significant cost savings if hazard strikes and, more importantly, it would result in much less human suffering.

These are but two small examples of the much larger question this book seeks to address:

What governance approaches and tools may help to improve the resource sustainability of our buildings and cities, may help to reduce their negative impacts on the natural environment and may make them more resilient to man-made and natural hazards?

In seeking to answer this question, I draw on research into urban sustainability and resilience from a ten-year period. I am an architect by training (MSc) and hold a PhD in public administration. Since 2005 I have been studying socio-technical and socio-legal aspects of the built environment. That is, I am interested in technological and legal solutions that may help to improve environmental and resource sustainability and the resilience of our buildings and cities. But I feel that 'just' technological and 'just' legal solutions are not sufficient.

Among others, I am interested in better understanding why very promising technological solutions for energy reductions have not been taken up on a large scale, and whether a sustainable and resilient built environment can be achieved with a minimal force of law. Over the last decade I have published widely on this topic in leading academic journals, and in journals and books for a policy and practitioner audience. I have spent time teaching on this topic and researching it further at various universities in Europe, Australia, the United States and India. You will find a more detailed biography in the preliminary section of this book.

In this book I bring together this ten years of experience, research and knowledge. I build on a series of interviews from various research projects with over 500 policy makers, public servants, administrators, architects, engineers, developers, investors, contractors, quality and quantity assessors, academics and other representatives of the building industry from around the globe. These individuals each spent more than an hour answering various sets of research questions I had prepared for the different research projects. In addition, they have provided me with a wealth of additional information. They have directed me to examples of governance for urban sustainability and resilience that I could not have found in the literature or through internet searches. They have shared anecdotes that provided me with richer insights into the complexities of governing urban sustainability and resilience. Lastly, they have given me brochures, news clippings, printouts of policy briefs, annual reports and addresses to relevant web pages that have further enriched my dataset. In this book I draw on all of these.

WHY I WROTE THIS BOOK

The idea to write this book emerged during my field trips. In my interviews I often find that my interlocutors are highly interested in what is going on in terms of governance for urban sustainability and resilience in other countries and in the academic literature. Time and again I have been told 'you should write a book about this!' This book follows up on their requests.

This book is unique in its scope. There now is a growing literature that discusses particular governance tools in urban settings, or the governance of urban sustainability and resilience in a particular area, such as energy savings, more generally (see Chapter 1). I am, however, not aware of any publication that maps, describes and evaluates such a wide set of governance tools for urban sustainability and resilience as I do in this book: close to 70 tools from Australia, Asia, Europe and North America.

In this book I only scratch the surface of all activity that is going on in the field of urban sustainability and resilience, and by no means do I claim that my book is representative for all this activity. I do, however, feel that the book provides a window on the opportunities and constraints that governments and non-governmental organizations face in governing urban sustainability and resilience.

WHAT YOU WILL GET FROM THIS BOOK

The current literature on governance, urban sustainability and resilience is vast and rich. In this book I provide a simple structure to better understand three main approaches to governance, and the tools that are available to governments, business and civil society groups and individuals to govern urban sustainability and resilience.

Each of the main chapters of this book addresses a particular approach to governance:

- Direct regulatory interventions in Chapter 2, which discusses the traditional tools governments have been applying for a long time. These are direct regulation, subsidies and other market interventions. I also discuss novel applications of these tools.
- Collaborative governance in Chapter 3, which addresses how governments, businesses and civil society groups are working more and more together in developing governance tools. Such tools are networks, covenants and (negotiated) agreements.
- Voluntary programmes and market-driven governance in Chapter 4, which has a focus on the frontiers in governance tools for urban sustainability and resilience. These are best-of-class benchmarking and certification, tripartite financing, green leasing, contests and challenges, and sustainable procurement.

These chapters are introduced by a discussion of the relevant governance literature, which is then illustrated with real-world governance tools that address urban sustainability and resilience. In total I discuss close to 70 different real-world governance tools in this book.

The theoretical structure of these chapters will be of interest to scholars who work on the edge of urban planning and governance and are interested in the opportunities that particular governance tools provide in achieving urban sustainability and resilience. The wide range of governance tools will be of interest to practitioners and policy makers in this field.

In Chapter 5 I bring this wide range of real-world examples together and discuss the five main trends in contemporary governance for urban sustainability and resilience. From here on I present 12 hands-on policy lessons. In Chapter 6, finally, I look towards the future.

In summary, this book will help you to formulate governance for urban sustainability and resilience that has a chance of successful outcomes. It will help you to develop governance tools that will likely work for the problems you seek to address. The broad knowledge base and the wide range of real-world examples of governance tools will help you to understand why well-trialled solutions often fail in improving the resource sustainability of our buildings and cities, in reducing their negative impacts on the natural environment and in making them more resilient to man-made and natural hazards. This book will stimulate and inspire you to think beyond such well-trialled governance tools for urban sustainability and resilience.

Acknowledgements

I could not have written this book without the support of the more than 500 individuals I have interviewed in recent years, comprising policy makers, administrators, architects, engineers, developers, financiers and users of buildings. I wish to thank them for their time and patience to answer my questions and provide me, often, with much more information than I could have hoped for. I am also grateful to Peter May and Ernst ten Heuvelhof for commenting on a first-draft outline of the book. Later in the book writing process I received invaluable support from Andy Jordan and Matt Hoffman. Their recent books, *Environmental Governance in Europe* (Wurzel et al., 2013) and *Climate Change at the Crossroads* (Hoffmann, 2011), respectively, have been a great source of inspiration. My colleagues at the Regulatory Institutions Network (RegNet, Australian National University) have all supported me tremendously in writing this book, but I am especially grateful to Veronica Taylor, Neil Gunningham, Peter Grabosky, Peter Drahos, Natasha Tusikov, Robyn Holder and Shane Chalmers. Thanks go to Gary Smailes from BubbleCow for helpful suggestions about how to improve the readability of the book.

I am particularly in debt to the encouragement, patience and support I received from Joost and Joachim throughout the ten-year journey that resulted in this book. To Gwen for supporting me on the first and most formative steps of this journey, and to Henrietta and Karissa for supporting me during the Australian part of it. I am also in debt to the Netherlands Organisation for Scientific Research that has generously supported me through an early career researcher's grant since 2012 (grant number 451-11-015).

Some aspects of Chapter 1 advances ideas that I introduced in Van der Heijden (2013c, 2013e, 2013f).

Chapter 2 strongly builds on my PhD thesis that was published as Van der Heijden (2009). Chapter 2 also incorporates ideas that I have developed and explored in Van der Heijden et al. (2007), Van der Heijden and De Jong (2009), Van der Heijden (2010a, 2010b, 2011), Vermande and Van der Heijden (2011) and Van der Heijden and Van Bueren (2013). Finally, Chapters 3 and 4 are inspired by and build on work that resulted from the research funded by the Netherlands Organisation for Scientific Research: Van der Heijden (2012, 2013b, 2013d, forthcoming 2014a, forthcoming 2014b, forthcoming 2014c).

1. Where we are today

In our urbanized world, cities and other urban environments[1] create problems as well as provide solutions. Cities are an unsustainable source of resource depletion and pollution, and account for 40 per cent of global energy consumption and over 30 per cent of greenhouse gas emissions. More than half of the world's population lives in ever-growing cities. Rapid industrialization in developing economies, ongoing urbanization and population growth will exacerbate these problems, while growing population density within cities itself breeds human and environmental stress and multiples the impact when disaster strikes. But there is hope.

Globally, governments, non-governmental organizations (NGOs) and businesses increasingly recognize that cities need to become more sustainable by reducing their impact on the natural environment, while becoming more resilient to natural and human-made hazards.[2]

But, how can we achieve cities that are environmentally and resource sustainable and resilient to human-made and natural hazards?

Some will argue that we can make significant improvements through technological innovations. Others will say that we can do so by rethinking our behaviour and changing the way we use cities and other urban environments. The technology and social know-how is indeed available to facilitate a cost-effective transition towards cities that are less dependent on energy, water and other resources, produce fewer greenhouse gases and other wastes and can better withstand natural or human-made hazards.

In other words, cities hold significant potential for increased sustainability and resilience. That is hopeful in itself. However, this technology and social know-how has been available for some decades. Still, most urban environments around the globe are far from being truly environmentally and resource sustainable and resilient to hazards.

This is the starting point of this book. It considers that technology and social know-how are necessary but not sufficient to achieve meaningful sustainable and resilient cities and other urban environments. The scale and speed with which this technology and social know-how is taken up is out of sync with the problems faced. This book argues that the governance of urban sustainability and resilience is another necessary (but not sufficient) part of the puzzle.

1.1 THREE MAIN GOVERNANCE PROBLEMS THAT HAMPER MEANINGFUL URBAN SUSTAINABILITY AND RESILIENCE

A well-known approach for seeking to achieve urban sustainability and resilience is the introduction of regulation by governments. Although such direct regulatory interventions have proved successful in the past (McManus, 2005; Otto-Zimmermann, 2010), the introduction of new regulation has significant drawbacks for improving the sustainability and resilience of today's cities and those of the future.

First, city governments in rapidly developing economies cannot respond fast enough to address current development rates of between 5 per cent to 10 per cent in terms of new buildings and infrastructure per year (IEA, 2009).

Not only do they lack regulatory frameworks capable of planning and delivering urban sustainability and resilience, they also lack robust data-based analyses of what is likely to work (Managan et al., 2012). Experience suggests that governments in rapidly developing economies can take decades to develop, implement and enforce sufficient regulatory requirements (Hong and Laurenzi, 2007; World Bank, 2009), resulting in a lag between urban development and regulatory response that locks cities into resource inefficiencies, high levels of greenhouse gas emissions and weak urban resilience (Bettencourt and West, 2010; Rosenzweig et al., 2010).

Second, city governments in developed economies face a different lag. Their cities develop and transform too slowly for new regulation to be meaningful.

Existing buildings and infrastructures are normally exempted from new regulation, a process known as 'grandfathering' (Nash and Revesz, 2007). Because most cities in developed economies develop and transform at less than 2 per cent per year it may take 40 to 70 years for new regulations to transform all buildings and infrastructure in cities in developed economies (IEA, 2009). Grandfathering in effect allows for the maintenance of weak links in the chain of sustainability and resilience in these cities. To give an illustration, as a result of grandfathering existing buildings, 80 per cent of existing Australian houses are exempted from compliance with building regulations introduced since the 1980s (Yates and Bergin, 2009).

Third, it becomes increasingly clear that an increase of urban sustainability and resilience provides for significant economic opportunities (IPCC, 2014, chapter 8).

Consumers are willing to pay higher prices for sustainable buildings (Eichholtz et al., 2010), while more resilient buildings may result in less financial harm when a natural or man-made hazard strikes (Australian Resilience Taskforce, 2013). In other words, there is a significant market potential for increased urban sustainability and resilience. Unfortunately, a wicked set of market barriers stands in the way for market players to exploit the (economic) opportunities that a transition towards more sustainable and resilient cities holds. These are the passing on of responsibilities for taking action, conflicting interests between various parties in the building industry and split incentives between those who pay for the costs of making the transition and those who gain from it. This wicked set of market barriers makes for a third governance problem that stands in the way for achieving meaningful urban sustainability and resilience.

To summarize, the three governance problems faced are:

- Governments are slow to react to existing problems of urban sustainability and resilience. It often takes a long time to develop and implement legislation and regulation, and even longer for these to cause their effects.
- Introducing new legislation and regulation is often inconsequential. In developed economies cities develop too slowly for new legislation and regulation to be meaningful. In developing economies cities develop too rapidly for new legislation and regulation to be meaningful.
- A number of market barriers stand in the way to capitalize the economic benefits that more sustainable and resilient cities can bring.

1.2 AIM OF THE BOOK

These three governance problems have focal attention throughout the book. This book seeks to understand what governance approaches and tools may help to overcome these problems.

Over the last decades governments have trialled innovative governance tools to address these problems, and businesses and civil society groups have become actively involved in governing urban sustainability and resilience. Solutions to the governance problems introduced are sought on international, supranational, national, regional and local scales, with especially city governments being actively involved in these (Hoffmann, 2011).

Throughout this book I demonstrate how the centre of gravity of governing for urban sustainability and resilience has, since roughly the 1990s, shifted from governments as sole-governance authorities to collaborations of governments, businesses and civil society groups; and to initiatives by businesses and civil society groups with limited involvement of government, if any. I further demonstrate how in these collaborations and initiatives city governments, in particular, have taken up new roles in governing urban sustainability and resilience. I finally present an overview of the breadth and depth of traditional and innovative governance tools that have resulted from these collaborations.

More specifically, in this book I seek to answer the question:

What governance approaches and tools may help to improve the resource sustainability of our buildings and cities, may help to reduce their negative impacts on the natural environment and may make them more resilient to man-made and natural hazards?

1.3 SUSTAINABILITY, RESILIENCE AND GOVERNANCE: SOME TERMS EXPLAINED

Urban sustainability, urban resilience and governance are popular buzzwords. In order to prevent misunderstanding each term is explained below.

1.3.1 Urban Sustainability and Urban Resilience

A classical understanding of urban resilience relates to how well cities are able to 'rebuild [their] physical fabric' after a disaster (Campanella, 2006, p. 141) and how well cities are able to 'maintain function … when shocked' (Rose, 2007, p. 384).

A clear definition or similarly confined understanding for urban sustainability is less easy to pin down (MacLaren, 1996; Wheeler and Beatley, 2009). Definitions of urban sustainability often build on aspects such as the protection of the natural environment and ensuring social equity, while allowing the economy to flourish within the boundaries of these (MacLaren, 1996; Wheeler and Beatley, 2009).

Agreeing that urban sustainability is a broad concept, in this book it is nevertheless addressed in a narrow understanding and relates to how urban environments can be developed, operated and maintained in such a way that they cause the least possible harm to the natural environment, predominantly through a more sustainable use of resources and a reduced production of wastes, greenhouse gases included. As such:

- Resilience may be considered a descriptive concept that gives insight into the particular properties of a city that make it capable to maintain functioning and recovering from disaster (UN, 2007).
- Sustainability, then, is a more normative concept that gives insight into a desired state of a city, and in this book, in particular, a minimal impact of that city on the natural environment through increased efficiency in resource use (Derissen et al., 2011).

Interestingly, the term 'urban resilience' appears to be slowly replacing and encapsulating the term 'urban sustainability', or at least the two terms appear to be used interchangeably by policy makers and practitioners.

The recent New York City Plan, with the progressive title *A Stronger, More Resilient New York*, is a case in point. On its very first page it defines resilience as: '1. Able to bounce back after change or adversity. 2. Capable for preparing for, responding to, and recovering from difficult conditions' (City of New York, 2013, inner cover page). Introducing the city plan to the people of New York, former Mayor Bloomberg lauds this plan as 'a roadmap for producing a truly *sustainable* 21st century New York' (City of New York, 2013, foreword from the mayor, emphasis added).

Another case in point is a recent publication by The Energy and Resources Institute (TERI), one of the Indian Government's prime advisory organizations in the area of urban development. In their publication *Mainstreaming Urban Resilience: Planning in Indian Cities – A Policy Perspective* the major challenge of resilience of cities is spelled out as 'to maintain environmental *sustainability*' (TERI, 2011, p. 2, emphasis added).

There is, of course, a close link between sustainability and resilience of urban environments. If cities become more sustainable, for instance, by applying more technologies and planning mechanisms that ultimately result in less consumption of energy, water and other resources, then they may be expected to be more resilient when disaster strikes. After all, less water, less energy and less other resources will then be needed to help the city to maintain functioning (for example, Folke et al., 2002; Milman and Short, 2008).

It is now becoming common to consider sustainability and resilience side by side, since there are so many synergies between them, particularly in urban environments (Ahern, 2013; Leichenko, 2011; Romero-Lankao and Dodman, 2011; Rosenzweig et al., 2010). Especially in terms of governing the transition *to* urban resilience, policy makers, practitioners and academics may very well be able to learn from the successes and failures of governing the transition to urban sustainability, and vice versa.

However, in urban settings there is some value in considering the two terms as referring to different outcomes. After all, a city may be highly resilient if it consumes large amounts of fossil fuels to maintain operation in a time of stress, but by doing so it is very unsustainable.

1.3.2 Governance

A well-known and well-trialled approach to governing urban sustainability and resilience is the introduction of mandatory regulation by (national) governments. More broadly, governance can be understood as an intended activity undertaken by one or more actors seeking to shape, regulate or attempt to control human behaviour in order to achieve a desired collective end (Dean, 2009; Foucault, 2009; Lemke, 2002).

All over the world governments have introduced construction codes that, for instance, stipulate how much energy a building should consume (that is, addressing urban sustainability) or how long a building should be able to withstand a fire (that is, addressing urban resilience). Yet, such direct governmental regulation is increasingly critiqued for being unable to achieve meaningful urban sustainability in a timely manner, and on a scale necessary to address the major problems posed by global change (these problems are further discussed in Section 1.5).

Governments on various levels, businesses and civil society groups and individuals have taken up the challenge to address these problems in ways that are different from traditional direct regulatory interventions. All over the globe governments are now participating in government-to-government networks such as the ICLEI (further discussed in Chapter 3), and share information and best practices about how to address urban sustainability and resilience in novel ways. Businesses and civil society groups have taken up the challenge by introducing their own regulations and governance programmes that seek to improve urban sustainability and resilience.

The Transition Town Network, for example, is a global citizen initiative that introduces regulation-like approaches to urban sustainability that move significantly beyond governmental regulation (further discussed in Chapter 4). Governments, businesses and civil society groups also collaborate to develop and implement governance tools that seek to achieve urban sustainability and resilience at levels that significantly surpass governmental regulation, but without the traditional force of law.

Typical examples are the best-of-class building benchmarking tools such as the BRE Environmental Assessment Method (BREEAM) and Leadership in Energy and Environmental Design (LEED) that have been mushrooming around the globe since the early 1990s. These tools stimulate developers and property owners to voluntarily build or retrofit buildings with higher levels of sustainable performance than required by governmental regulation, and to showcase the sustainability credentials of their buildings (further discussed in Chapters 3 and 4). Finally, supported by businesses and civil society groups, governments have taken up new roles in the governing of urban sustainability and resilience. Especially city governments appear to have become initiators, assemblers and supporters of innovative governance tools (further discussed in Chapters 3 and 4).

These innovative governance tools, the involvement of businesses and civil society groups in governing and the new roles of, particularly, city governments are very much in line with a wider trend in contemporary governance. This trend was, perhaps, best captured by Rod Rhodes who in the mid-1990s argued that there has been a shift 'from government to governance' (Rhodes, 1997, 2007).

Scholars of governance are often interested in the difference(s) between governing through direct regulatory and coercive interventions by governments ('old governance') and innovative approaches to governing in which multiple actors are involved ('new governance'). In their work they then address a 'who' question: Who are the 'one or more actors' that seek 'to shape, regulate or attempt to control human behaviour' and which actor is, or which actors are best suited to govern?

The work of these scholars shows that over the years the government has become less directly involved in governing and non-governmental actors take on much of the 'heavy lifting' of governance (Chhotray and Stoker, 2010; Kickert et al., 1997; Rhodes, 1997, 2007; Teisman and Klijn, 2002).

Scholars of governance further address 'what' and 'how' questions: What are the most suitable tools to shape, regulate or attempt to control human behaviour? How do these tools operate in real-world settings?

Scholars then seek to compare traditional governmental tools of government, such as direct regulation and subsidies, with contemporary governance tools. These questions are particularly important to address for the governing of urban sustainability and resilience because, as this book highlights, such a wide range of new governance tools have been introduced in this area. Academics, policy makers and practitioners have only begun to understand the importance of these innovative governance tools for making a transition to meaningful urban sustainability and resilience (Otto-Zimmermann, 2010).

This book follows this tradition of governance studies by asking these who, what and how questions for a large sample of governance tools, in a variety of settings, in which different governmental, business and civil society actors seek to achieve urban sustainability and resilience.

1.3.3 The Need to Govern Urban Sustainability and Resilience

Urban environments are considered to account for 40 per cent of total energy consumption (Pérez-Lombard et al., 2009). About 10 per cent of all potable water is used in urban environments, industry excluded. This may seem like a small number, but 90 per cent of this potable water is returned as waste water and is often in such a degraded state that it requires major treatment before it can be used again (World Water Council, 2007). Globally, cities are over-abstracting groundwater and other water sources, which has a negative impact on the (natural and rural) environment that surrounds cities (UN, 2012). Finally, the development of urban environments accounts for about 45 per cent of all extracted raw materials (Kibert, 2008), adding to ongoing deforestation, mining for metals and stone products, and the subtraction of other resources that are essential parts of other species' ecosystems.

But not only do urban environments have a major negative impact on the natural environment by extracting resources from it. The development and maintenance of urban environments is considered to produce about 40 per cent of all wastes (UNEP, 2003), many of which take a long time to degrade, such as plastics, asphalt or concrete. Urban environments are considered to account for 35 per cent of global greenhouse gas emissions, and this is excluding urban transport (Dodman, 2009; IPCC, 2014). This waste and pollution not only has a negative impact on the natural environment but results in significant health risks. For instance, many cities around the world were not designed (if designed at all) for the current pressure of transport they are facing. This overload of traffic causes dangerous levels of air pollution, with related negative health consequences (Gurjar et al., 2010; Pope et al., 2009). With ongoing urbanization, ongoing industrialization, ongoing economic development and a growing world population these problems are only expected to increase over the next decades (Alexander, 1997; Hochrainer and Mechler, 2011; Hong and Laurenzi, 2007; IPCC, 2014; Sen, 2013; Smith, 2013; UNEP, 2007).

To make things worse, because of ongoing urbanization natural and human-made disasters appear to strike harder than ever before. It is now expected that by 2050 the world population will have grown to 9.6 billion people and that 70 per cent of the world population will then live in cities and other urban environments (UN-HABITAT, 2013).

With more people living in urban environments, and with cities increasingly becoming the centres of economic development, urban disasters appear unavoidable. Hurricane Sandy, which hit New York on 29 October 2012 (after causing mayhem in the Caribbean) and Hurricane Katrina, which hit New Orleans on 29 August 2005, are two typical examples. Both hurricanes caused large numbers of casualties and significant economic damage in a country that is normally not considered to be prone to extreme weather events with devastating consequences (Robertson et al., 2007; Schmeltz et al., 2013). The Fukushima Daiichi nuclear incident on 11 March 2011 showed how quickly a human-made risk can evolve to a disaster, while the nationwide power blackouts in India on 30 and 31 July 2012 give an illustration of the impacts of a slowly evolving human-made disaster that has a major impact on, especially, urban environments (Romero, 2012; Ten Hoeve and Jacobson, 2012).

Such disasters have spurred action in this field. Interestingly, not only governments and (international) NGOs are becoming interested in improving urban sustainability and resilience. Major companies have recently started to map the risks for urban environments and now work together with governmental and non-governmental organizations to seek solutions to mitigate these risks (Arup et al., 2013; Rockefeller Foundation, 2013; Swiss RE, 2013). An example that I discuss in Chapter 3 is the SMART 2020: Cities and Regions Initiative. An initiative by an international developer of networking equipment, and the cities of Amsterdam, San Francisco, Seoul, Hamburg, Lisbon and Madrid. The initiative seeks to understand how current and innovative information and communications technology (ICT) applications can help to reduce the greenhouse gas emissions in cities.

1.3.4 Technological Fixes, a Partial Solution

It has been argued that we have, or will soon get, a technological fix for the urban challenges caused by global changes such as urbanization, a growing world population and the rapid development of economies in the global South (for example, Register, 1987; Roodman and Lenssen, 1995; Roseland, 1997; Stren et al., 1992).

Indeed, in terms of reducing cities' impact on the natural environment major advancements have been made. The use of renewable energy technologies is possibly the most important of these, and in particular the use of solar energy through photovoltaic technologies in urban settings (Droege, 2008; Parida et al., 2011; but see the German example discussed in Chapter 2).

Significant steps have been made in developing new construction materials that reduce the energy required for heating or cooling buildings, for instance, double or triple glazing (Jelle et al., 2012; Sadineni et al., 2011). In terms of reduced water consumption, much has been achieved in rainwater harvesting for use inside buildings (Domènech and Saurí, 2011; Rahman et al., 2010), the reuse of potable water by, for instance, flushing toilets with water earlier used for showering (Mourada et al., 2011) or replacing ordinary toilets with toilets that do not consume any water at all (Anand and Aspul, 2011). Keeping in mind that toilets are one of the major sources of water consumption in urban environments (up to 60 per cent of all building related water use in developed economies), it goes without saying that such small changes can have a major overall impact (McAllister and Sweett, 2007; Otterpohl and Buzie, 2011). In terms of waste reduction and a reduced use of raw-materials, progression has been made in (partly) recycled and reused construction materials such as recycled course aggregate (often from concrete from demolished buildings) in new concrete (Kwah et al., 2012), or the use of corn cob (an agriculture waste) as insulation material for buildings (Pinto et al., 2012).

But not only such 'high-tech' solutions are considered an answer to the current lack of sustainability of urban environments. Much may be achieved in terms of urban form (Echenique et al., 2012; Hamin and Gurran, 2009).

For instance, cities may be designed in ways that reduce (car) transport miles for their inhabitants (Glaeser and Kahn, 2010; Graham-Rowe et al., 2011), which make optimal use of roofs as areas for food production (Peters, 2010; Specht et al., 2013) or that seek higher densities for their inhabitants to reduce urban sprawl (Bart, 2010; Cadigan et al., 2011). Building and infrastructure designers are becoming more aware of relatively simple solutions, such as making optimal use of a building's orientation to the sun (Morrisey et al., 2011; Z. Yu et al., 2013); and with simple modifications, such as painting an existing building's rooftop white, significant energy savings may be achieved because the building accumulates less heat (Jo et al., 2010; Scherba et al., 2011).

In addition, by using various combinations of these technologies and solutions, it is now possible to develop buildings that are energy and water passive and do not rely on traditional utilities for these resources (Englehardt et al., 2013; Robert and Kummert, 2012). It is now even possible to develop buildings that produce more energy than they use (Kolokosta et al., 2011). The surplus of energy generated may, for instance, be used to power electric vehicles, which may ultimately lead to cities that do not produce any carbon emissions at all, that is, net-zero carbon cities (Novotny, 2012).

Addressing the impact of human-made and natural hazards is a well-explored topic in the technical sciences. Over the years (and decades), much progress has been made in reducing cities' vulnerability to natural hazards by protecting cities from flooding (Feynen et al., 2009; Gandy, 2008), heat waves (Depietri et al., 2012; Jenerette et al., 2011) and storms (Blocken et al., 2012; Ma et al., 2013). Buildings and other infrastructure can now be designed and developed in such a way that they are highly fire-resistant (Hasofer et al., 2007; Lankao and Qin, 2011) and earthquake-proof (Albrito, 2012; Takewaki et al., 2011). Potential hazards can now be identified before they escalate, and aid can be delivered faster and more targeted if a catastrophe occurs, allowing for cities to maintain functioning when disaster strikes (Asimakopoulou and Bessis, 2010; Zhang et al., 2013). Finally, application of the discussed 'sustainability' technology holds the promise of cities that are self-sufficient, which is an important aspect of resilience.

To summarize, there is an abundance of traditional and innovative technology that may help to reduce resource consumption in cities, reduce their negative impact on the natural environment and may help to make cities more resilient. However, while this technology has been moving forward for decades its implementation lacks in both speed and scale to be meaningful in addressing the problems of urban unsustainability and non-resilience. Technology is necessary but not sufficient to address these problems.

1.3.5 Behavioural Change, Another Partial Solution

Technology is but a means to an end and not an end in itself. It has to be utilized to have an effect. An increasing amount of knowledge is generated in regards to how a changed use of an existing technology, and a changed use of existing buildings and cities may be key in addressing the urban challenges posed by global change.

Scholars sometimes point towards a resistance of policy makers, businesses, practitioners and citizens to use the innovative technological innovations discussed before (Hoffman and Henn, 2009; Revell and Blackburn, 2007; Sengers et al., 2010; Smith et al., 2005).

Practitioners may be opposed to using certain technological innovations when they question the performance of the new technology versus the business-as-usual technology. The use of recycled coarse aggregate in concrete is a typical example of such an innovation that has faced (and faces) much resistance from practitioners. For a prolonged period contractors have not trusted the quality and strength of this material as being at par with conventional concrete (Kwah et al., 2012; Nixon, 1976).

Another issue that stands out is vested interests (Jacobsson and Bergek, 2011; Moe, 2012). Producers and suppliers of business-as-usual technology may lobby against the development or use of new technologies simply because they fear competition with their own technologies.

Finally, policy makers may, for good or bad reasons, be against the application of new insights. An illustrative example here is the recycling of waste water. By the turn of the millennium building practitioners and the scientific community alike had shown that grey water and waste water could safely be recycled and (re)used in domestic settings (Jefferson et al., 2000). In many countries, however, building codes or health regulations do not allow for the use of such water, or the instalment of grey water systems. An example comes from the City of Oakland, CA, United States. The Building Code of California set very restrictive specifications for grey water systems, which in practice made them too expensive to install.

Yet, many of those wishing to install a grey water system do so hoping to use the recycled water for watering their gardens or washing their cars, and not for (human) consumption (Brevetti, 2008). Aiming to get this regulatory issue addressed by legislators, a group of educators, designers, builders and artists have come together to educate the public and to give voice to those wishing for a more sustainable water culture and infrastructure. They have been doing so since 1999, for ten years under the name Greywater Guerrillas and since 2009 as Greywater Action. The group achieved some of their goals with the implementation of the progressive 2009 California Greywater Code, which allows for the instalment of grey water systems and even encourages the use of grey water (California Department of HDC, 2009).

Other perceptive insights into how urban environments may become more sustainable and more resilient come from the behavioural sciences and, related, behavioural economics (Amir and Lobel, 2008; Heiskanen et al., 2010; Osbaldiston and Schott, 2012; Timmer, 2012). Research from these fields show that with small behavioural changes huge effects may be achieved.

A typical example is 'phantom power', the power consumed by electronic devices and appliances when they are switched off or in standby mode (Chakraborty and Pfaelzer, 2011; Popovic-Gerber et al., 2012; Rusk et al., 2011). About 10 per cent of electricity consumption in households (in developed economies) comes in the form of phantom power (IEA, 2001), while recent research indicates that about 55 per cent of the energy of Australian office buildings is consumed after office hours, during weekends and holidays when no one is using these buildings (Greensense, 2013). Households and office users are often not aware of such wastes or do not know how to address them. Awareness campaigns, the labelling of their electronic appliances or supporting households and office managers to use power cut-off systems may help to reduce this unnecessary loss of energy (De Almeida et al., 2011; Solanki et al., 2013).

Alternatively, providing insights into comparative usage can see household and industry energy consumption drop. Some research points in the direction that households and firms will reduce their energy consumption when shown that they consume more than the average household or firm, for instance, their neighbour or competitor (Allcott, 2011; Davis, 2011). Creating, changing and stressing social norms about issues such as energy and water consumption, carpooling or recycling may very well be as important to achieve urban sustainability and resilience as creating even more efficient technological solutions.

To summarize, over the last decades great advantages have been made in our understanding of how people use buildings and cities. Resource use can be reduced significantly simply by rethinking how we use our buildings and cities. However, acting to these insights about behavioural change falls short in both speed and scale to be meaningful. Like technology, behavioural change is a necessary but not sufficient condition to address the problems of urban unsustainability and urban non-resilience.

1.3.6 The Missing Part of the Puzzle: Getting Governance Right

There appear to be no limits as to what technology and behavioural change *may* bring in terms of improving the sustainability and resilience of urban environments. Also, with ongoing technological innovation and increased availability of technology it becomes both cheaper and easier to reduce energy, water and raw material consumption and to make cities less dependent on the external provision of these resources (Branker et al., 2011; Hoffman and Henn, 2009; Lubin and Esty, 2010; Nill and Kemp, 2009). The construction, operation and maintenance of urban environments are considered to hold the highest potential for cost-effective greenhouse gas emission reductions (IPCC, 2014).

With current technology and social know-how, cities can be built and operated in such a way that they are less harmful for the natural environment and are less affected by natural or human-made hazards, while at the same time saving resources and costs. Cities have a large-scale potential where many small changes in individual buildings and infrastructure may together add up to a significant transformation.

In short, 'the [construction, operation and maintenance of buildings] has more potential to deliver quick, deep and cost-effective GHG [greenhouse gas] mitigation than any other [sector] … [and] can be achieved in the short term' (UNEP, 2010, p. 2; see also ICCP, 2014, chapter 8).

Still, after decades, or even more than a century (for example, Howard, 1902), of development of this technology and social know-how the real world does not show large-scale and speedy uptake of it (for example, UN-HABITAT, 2008, 2009). Technology and behavioural change are necessary to improve the sustainability and resilience of urban environments, but they are not sufficient in addressing the urban challenges posed by climate change and a growing world population.

This brings us full circle to the discussion on governance. Addressing the governing of a transition towards meaningful urban sustainability and resilience raises a wide range of questions.

- What actor or actors are best positioned to stimulate a wider uptake of the available technology and social know-how?
- Should governments demand the use of this technology or implement regulation that coerces people into desired behaviour; and if so, at what level should government operate?
- What are promising governance tools for government to implement?
- Can we expect the market to take it up; and if so, what are effective market governance tools?
- Is there perhaps an in-between approach between government and the market?

- And what are effective governance tools for this in-between approach to governing?

Questions like these are, as earlier discussed, at the core of many governance studies. To answer them for urban sustainability and resilience it is of importance to understand the specific governance problems faced in this area.

1.4 THE GOVERNANCE PUZZLE FOR URBAN SUSTAINABILITY AND RESILIENCE: OVERCOMING THREE MAJOR GOVERNANCE PROBLEMS

Three major governance problems stand in the way of governing cities towards urban sustainability and resilience. I have already briefly touched on these in the opening pages of this book, but some more explanation may be necessary.

1.4.1 The Problem of Grandfathering

Grandfathering refers to the exempting of existing buildings and infrastructure from new or amended regulation (Nash and Revesz, 2007; Shavell, 2007; Vinagre Diaz et al., 2013). Buildings and (privately owned) infrastructure come with a set of property rights that relate to a certain time period and are very difficult to change. Existing buildings and infrastructure make up the large majority of cities (about 98 per cent; see, IEA, 2009; McAllister and Sweett, 2007). In addition, existing buildings are replaced only very slowly (once every 50–70 years; see, Balaras et al., 2007; Fay et al., 2000).

Generally speaking, the older the buildings or infrastructure, the less sustainable and resilient they are. Existing buildings in developed economies, for example, have a greenhouse gas emission savings potential of 40–60 per cent (IEA, 2009). Yet, governing urban sustainability and resilience by introducing new regulatory requirements will unlikely result in meaningful sustainable and resilient cities in, especially, developed economies because only 2 per cent of new buildings are added to existing cities in these countries per year (IEA, 2009). In effect, grandfathering allows for the maintenance of weak links in the chain of urban sustainability and resilience.

Phrased differently, in developed economies cities develop and transform too slowly for direct regulatory interventions to have a meaningful effect on improving urban sustainability and resilience.

1.4.2 The Problem of Regulating a Building Boom in Developing Economies

Since the turn of the millennium, developing economies have been facing an unprecedented building boom. The building stock in India, for instance, doubled between 2002 and 2007 (Hong and Laurenzi, 2007). Since the early 2000s, developing economies more generally have seen their building stock grow by 5–10 per cent per year (IEA, 2009).

Addressing new buildings and new infrastructure is of major importance in developing economies to achieve sustainable and resilient cities. This may prevent these cities becoming locked in to similar energy and water inefficiencies as cities in developed economies. It is therefore of importance for governments in developing economies to implement regulatory requirements that mandate the construction of buildings and infrastructure that are highly sustainable and resilient. However, these developing economies particularly face significant governance, knowledge and financial barriers for doing so (Hong and Laurenzi, 2007).

There is a reported lack of knowledge on information about solutions that may result in sustainable and resilient cities, a lack of willingness in the building sector to use new solutions across the board and a lack of willingness at policy level to make necessary changes (Hong and Laurenzi, 2007).

In addition, introducing new regulatory requirements is by no means a guarantee for achieving desired outcomes. When introduced by governments, such requirements need to be enforced. Again, especially in developing economies, such enforcement is often complicated by a reported lack of enforcement capacity at government level, or a lack of willingness of regulators and politicians to take necessary enforcement actions (Hettige et al., 1996; Kirkpatrick and Parkers, 2004; Nath and Behera, 2011).

In summary, because of economic and institutional circumstances the governing of urban sustainability and resilience in developing economies asks for different governance tools than direct regulatory interventions introduced and enforced by governments.

1.4.3 The Problem of a Wicked Set of Market Barriers

The transition towards urban sustainability and resilience has major economic potential. Reducing the consumption of energy, water and raw materials is possible, and can be done cost-effectively with current technology and social know-how (IPCC, 2014). Such reductions will make cities and other urban environments more sustainable. At the same time, many future costs may be prevented if existing buildings and infrastructure are made subject to current-day regulatory requirements, and if new buildings and infrastructure are built to the highest safety standards (ClimateWise, 2012; IIGCC, 2013). Buildings and infrastructure, both new and existing, are then more likely to resist natural or human-made hazards, and in conjunction will make cities more resilient.

With such an economic impact one would expect that market players want to take up this potential and start collaborating with governments to govern urban sustainability and resilience, or even start governing for urban sustainability and resilience on their own initiative. As will be discussed throughout the book, they do. However, a wicked set of intertwined market barriers provides for another complex governance problem, which hampers governance initiatives by market players.

In particular, the building sector is a notoriously conservative sector in which actors are not eager to change business-as-usual practice to more sustainable (and more resilient) practice (Beerepoot and Beerepoot, 2007; Rees, 2009).

Developers and (product) manufacturers in the sector face significant first-mover disadvantages, and have to bridge significant 'chasms' to bring their products and services to the market (for terminology, see Dobrev and Gotsopoulos, 2010; Moore, 2002). These relate to the financial, legislative and cultural risks organizations face when bringing a new product or service to the market. The new product or service may be considered too expensive by clients, may conflict with existing legislation or may face resistance when it is considered 'ahead of its time' or 'too fast for the market' (Robinson and Min, 2002).

A particular issue that stands out is the passing on of responsibilities by various actors in the sector, sometimes referred to as the 'vicious circle of blame' (Cadman, 2007). This vicious circle of blame refers to a situation in which all parties involved blame each other for not providing, demanding or financing buildings or infrastructure with high levels of environmental performance. The building sector is further characterized by a wide range of professions and trades, such as architects, engineers, technical advisers, contractors, developers, investors and property owners – some of these may operate on a local level, while others operate on an international level. They often have their own representative bodies, which lobby actively to see their interests served (Lillie and Greer, 2007). Also, split incentives between, for instance, landlords and tenants or current and future property owners mean that property owners often do not feel a need to improve the environmental sustainability of their property because they do not receive the financial gains from, for instance, reduced energy or water consumption (Gillingham et al., 2009; Sorrell et al., 2004).

On top of all this, every construction project is unique. As a result, the rule of economies of scale does not apply for solutions that may make buildings and infrastructure projects as a whole more sustainable or resilient.

In summary, although the transition to sustainable and resilient cities has major market potential, it is unlikely that governance initiatives by market players only will be successful in addressing the urban challenges posed by global change on the scale and speed required.

1.5 FOCUS AND OUTLINE OF THE BOOK

Throughout the book I discuss traditional and innovative governance tools that seek to facilitate a transition towards more sustainable and more resilient cities. In particular, I focus on governance tools that address buildings, both existing buildings and buildings that are still to be built or that are under construction ('new' buildings). I cluster these tools in three approaches to governance based on the actors involved in developing and implementing the tools.

1.5.1 A Focus on Three Approaches to Governance

In order to better understand governance for sustainable and resilient urban environments, the actors involved in it and the governance tools used, this book addresses three specific governance approaches: direct regulatory interventions; collaborative governance; voluntary programmes and market-driven governance.

This clustering is for heuristic purposes only. It is a relatively easy way to distinguish among governance tools based on the dominant actors involved in their development and implementation.

The cluster direct regulatory interventions addresses tools developed and implemented by predominantly governmental actors such as:

- statutory regulation
- direct subsidies
- taxes
- other economic instruments.

The cluster collaborative governance addresses tools developed and implemented by alliances of governments, NGOs, businesses and citizen groups such as:

- networks
- partnerships
- agreements and covenants.

The cluster voluntary programmes and market-driven governance, finally, addresses tools developed and implemented predominantly by non-governmental actors with governments at some distance such as:

- best-of-class benchmarking and certification
- tripartite financing
- green leasing
- contests and challenges
- sustainable procurement.

The clustering will likely feel natural to policy makers and practitioners, and resembles a fairly accepted way of clustering approaches to governance in the academic literature (for example, Baldwin et al., 2011; Delmas and Young, 2009; Price and Verhulst, 2005; Tommel and Verdun, 2009; Williamson, 1996).

1.5.2 A Focus on New and Existing Buildings

The remainder of this book predominantly focuses on real-world examples of governance tools that aim for a transition towards buildings (and cities) that are more sustainable and more resilient than contemporary buildings.

I am aware and agree, as illustrated before, that urban sustainability and resilience encompasses more than environmental and resource sustainable, and hazard resilient buildings. I do however use the 'case' of governance for sustainable and resilient buildings as an illustration for the broader topic of urban sustainability and resilience.

A number of reasons support my choice.

First, buildings are key for urban environments, since without buildings, no urban environment can exist. And, as per the urban challenges posed by global change, the lion's share of energy consumption, water consumption, raw material consumption, greenhouse gas emissions and other waste production in urban environments, is taken up by the construction, operation and maintenance of buildings. Related to this, it is in the construction, operation and maintenance of buildings that most savings and reductions may be achieved (IPCC, 2014). In short, buildings are particularly important in achieving urban sustainability and resilience.

Second, buildings have for long been subject to governmental regulation and zoning laws – see, for instance, the development of early day building regulations during the industrial revolution in England (Ash and Ash, 1899; Emden, 1885), the United States (Gould, 1895), the Netherlands (De Ranitz, 1948; De Vreeze, 1993; Kocken, 2004) and France (Risler, 1915). From the nineteenth century on, this regulation has been adapted to suit contemporary needs and, worldwide, present day building regulation covers a broad range of topics, such as safety, public health, amenity and sustainability – see, for instance, present day building regulations in Australia (ABCB, 2004), different European countries (Van der Heijden and Van Bueren, 2013), India (BIS, 2005), Malaysia (Department of Standards Malaysia, 2007), Singapore (BCA, 2008) and the United States (ICC, 2006). As illustrated, there appear to be limits to what direct regulatory interventions may be able to achieve in addressing urban sustainability and resilience.

Third, for some two decades now, businesses, NGOs and citizen groups have been active, particularly on a local level, to increase aspects of buildings' sustainability. All over the world, buildings have been subject to a range of less coercive governance tools than direct regulatory interventions (Beatley, 2000, 2009; Evans et al., 2005; Hickson, 2009; Moore, 2007; Newman et al., 2009). These may provide important lessons on what governance tools may cause positive outcomes in overcoming the wicked set of market barriers discussed, and by doing so in achieving sustainable and resilient cities and other urban environments.

1.5.3 A Focus on Real-world Examples of Governance for Urban Sustainability and Resilience

This book builds on a series of research projects that I have carried out between 2005 and 2013. All projects sought to better understand the governance of urban affairs such as urban sustainability (Van der Heijden, 2013a, 2013b, 2013d, forthcoming 2014a, forthcoming 2014b, forthcoming 2014c), energy performance of buildings in Europe (Van der

Heijden and Van Bueren, 2013; Vermande and Van der Heijden, 2011)[3] and compliance with building codes (Van der Heijden, 2009, 2010a, 2010b, 2013a). These projects are built on data collected in Australia, Canada, Germany, India, Malaysia, the Netherlands, Singapore, the United Kingdom and the United States.

My major source of data are interviews with over 500 experts[4] in the building sector, such as architects, engineers, contractors, developers and financers, as well as policy makers, policy administrators and academics. For this book I have supplemented this dataset with data from existing academic and 'grey' literature, such as research reports, websites and newspaper articles. Table 1.1 provides an overview of the real-world examples of governance tools for urban sustainability and resilience that I discuss throughout the book.

Table 1.1 Real-world examples of governance tools addressed in this book

100 Resilient Cities Centennial Challenge	Network of cities selected by the Rockefeller Foundation. The cities receive technical and financial support for developing and implementing plans for urban resilience.
1200 Buildings	Melbourne-based tripartite financing tool that funds retrofits of existing commercial property.
Aldinga Arts Eco Village	Village in the state of South Australia, in which citizens have implemented far-reaching by-laws to achieve high levels of urban sustainability.
Amsterdam Investment Fund	Revolving loan fund of the City of Amsterdam that issues loans to, among others, building developments and retrofits that seek to achieve high levels of environmental performance.
Australian National Construction Code	Statutory building regulation in Australia (of particular interest for its combination of prescriptive and performance-based regulation).
Better Building Partnership	Partnership between the City of Sydney and local commercial property owners committed to reduce their energy consumption.
Billion Dollar Green Challenge	US-wide programme that encourages colleges, universities and other non-profit institutions to invest a combined total of US$ 1 billion in self-managed revolving funds that finance energy-efficiency improvements.
Bombay First	Collaboration of Mumbai-based businesses and governments committed to improve sustainability and resilience of the city.

BREEAM (BRE Environmental Assessment Method)	Best-of-class benchmarking tool for buildings. Aims to stimulate developers and property owners to build and retrofit buildings with high levels of environmental performance. Originally started in the United Kingdom, now applied throughout the world
Building Innovation Fund	Competitive grant by the Government of South Australia. It seeks to fund the most promising proposals for building retrofits in terms of urban sustainability.
Building Resilience Rating Tool	Tool developed by the Australian insurance industry that rates the resilience of homes to common extreme weather events.
C40 Cities Climate Leadership Group	Global network of 58 major cities that aims to develop and share lessons and best practices about urban sustainability and resilience.
Canadian Renewable and Conservation Expenses	Canadian tax deduction for the upfront costs of developing and exploring the application of renewable energy.
Capacity and Development Grant	Grant by the Government of Singapore to support small and medium-sized businesses to achieve ISO certification in areas such as environmental management.
Chicago Green Office Challenge	Chicago-based, within-office and office-to-office competition. Participants (office tenants) aim to improve their environmental sustainability.
CitySwitch Green Office	Australia-wide network in which office tenants are supported by local governments to improve their environmental sustainability.
Climate Change Sector Agreements	Negotiated agreements between the Government of South Australia and local businesses committed to reduce their greenhouse gas emissions.
ClimateSmart Home Service	Door-to-door programme by the state of Queensland to inform households about how to reduce energy and water consumption.
Common Carbon Metric	Metric aiming to make the carbon intensity of buildings around the globe comparable. Developed with support from the United Nations.
Density Bonuses North Carolina	A number of jurisdictions in North Carolina provide density bonuses to builders who build or retrofit energy-efficient buildings.
DGNB System (Deutsche Gütesiegel Nachhaltiges Bauen)	Best-of-class benchmarking tool for buildings. Aims to stimulate developers and property owners to build and retrofit buildings with high levels of environmental performance. Originally started in Germany, now applied throughout the world.
Dutch Energy Service Company contracting	Energy Service Companies (ESCOs) aim to reduce the energy consumption of their clients. The Dutch Government supported ESCOs in reducing the legal barriers for them to operate.

E+Green Building	Design competition by the Government of Boston. It sought the most promising designs for multi-unit housing that produce more energy than they consume.
Eco-Office	Best-of-class benchmarking tool for office tenants. Aims to improve the environmental sustainability of office tenants.
Energy End-use Efficiency and Energy Services Directive	European-wide directive that requires the 27 European Union member states to draw up national action plans to achieve 1 per cent annual energy savings over nine years starting in 2008.
Energy Experience Programme	Singapore educational programme. It aims to make secondary school students aware of their energy consumption, and that of their household.
Energy Performance Certificates	European-wide information tool, but implanted in different forms in different countries. The certificate highlights the energy performance of buildings.
Energy Performance of Buildings Directive (EPDB)	European-wide directive that requires the 27 European Union member states to set minimum energy performance requirements for residential and commercial buildings; and, to encourage beyond compliance behaviour.
Environmental Upgrade Agreements	Sydney-based tripartite financing tool that funds retrofits of existing commercial property.
ESCO Accreditation Scheme	Singapore programme that seeks to enhance the professionalism of ESCOs and the quality of services they deliver.
Feed-in-Law Germany	German law that guarantees fixed prices for energy generated from renewable resources for 20 years, and mandates grid operators to connect renewable energy installations to the net.
Future Cities Network	Collaborative of nine cities in the United Kingdom, the Netherlands, Belgium and Germany that seeks to make city regions in Northwest Europe fit to cope with predicted climate change impacts.
Grassroots Program	Competitive grant by the City of Boston that programmes the design and construction of community gardens and open spaces; and conveys city-owned land to non-profit organizations for community benefit and use.
Green Building Incentive Program	County of San Diego's incentive of reduced plan check turnaround time and a 7.5 per cent reduction in plan check and building permit fees for projects that have significantly higher levels of environmental performance than required by the legislation.
Green Building Index	Malaysian best-of-class benchmarking tool for buildings. Aims to stimulate developers and property owners to build and retrofit buildings with high levels of environmental performance.

Green Building Tax Credit Program	Tax credits for the construction and retrofitting of energy-efficient buildings in the state of Maryland.
Green Deals	Covenants between the Government of the Netherlands and local businesses and households committed to reduce their greenhouse gas emissions.
Green Door	State of Queensland incentive of reduced plan check turnaround time for building proposals that have significantly higher levels of environmental performance than required by the legislation.
Green Labelling Scheme	Best-of-class benchmarking tool for products and services. Aims to stimulate developers and property owners to use building products and services with high levels of environmental performance.
Green Leasing Toolkit	Website that brings together information about green leases. It helps organizations in California to develop green leases, communicates policies on urban sustainability to the market and seeks to develop a language for green leases.
Green Mark	Best-of-class benchmarking tool for buildings. Aims to stimulate developers and property owners to build and retrofit buildings with high levels of environmental performance. Originally started in Singapore, now applied throughout Asia.
Green Star	Best-of-class benchmarking tool for buildings. Aims to stimulate developers and property owners to build and retrofit buildings with high levels of environmental performance. Originally started in Australia, now applied in various countries, including South Africa.
Green Strata	Australian-wide programme that provides information to owners of units in multi-residential buildings and their managers on how to improve the environmental sustainability of these buildings.
GRIHA (Green Rating for Integrated Habitat Assessment)	Indian best-of-class benchmarking tool for buildings. Aims to stimulate developers and property owners to build and retrofit buildings with high levels of environmental performance.
Growing a Green Heart Together	City of Brisbane's community plan that was developed in close collaboration with the city's citizens. The plan aims for a carbon neutral city by 2026.
ICLEI – Local Governments for Sustainability, (International Council for Local Environmental Initiatives)	Global network of more than 1000 cities that aims to develop and share lessons and best practices about urban sustainability and resilience.
International Green Construction Code	Set of private standards, a code, that seeks to improve the environmental performance of buildings

LEED (Leadership in Energy and Environmental Design)	Best-of-class benchmarking tool for buildings. Aims to stimulate developers and property owners to build and retrofit buildings with high levels of environmental performance. Originally started in the United States, now applied throughout the world.
Local Law 86	The City of New York's commitment to have all governmental buildings LEED certified.
NABERS (National Australian Built Environment Rating System)	Australian best-of-class benchmarking tool for buildings. Aims to stimulate developers and property owners to build and retrofit buildings with high levels of environmental performance. Originally started as a voluntary programme, now compulsory for particular building types.
National Flood Insurance Program	Subsidized insurance to homeowners that live in flood-prone areas in the United States.
National Green Leasing Policy	Commitment of Australian governments to use green leases for their office space.
Ontario Regional Adaptation Collaborative	Collaboration by the Governments of Canada and Ontario that seeks to support communities to adapt to climate change.
Open Mumbai	Collaboration of a Mumbai-based architect, citizens and government that seeks to improve sustainability and liveability in the city.
PACE (Property Assessed Clean Energy)	Tripartite financing programme that allows local governments in the United States to issue bonds to investors and use these funds as loans for energy retrofits to homeowners and commercial property owners.
Peer Experience and Reflective Learning Network	Indian network of 65 cities that aims to develop and share lessons and best practices about urban sustainability and resilience.
Privatised Building Control Australia	Involvement of private sector certifiers in the enforcement of Australian construction codes.
Privatised Building Control Canada	Involvement of private sector certifiers in the enforcement of Canadian construction codes.
Public Procurement Bill 2012	Voluntary code on sustainable procurement in India. The bill stipulates that environmental criteria of a product or service may be adopted as one of the criteria for the evaluation of a tender by Indian governments.
SMART 2020: Cities and Regions Initiative	Collaboration by a major ICT firm and the cities of Amsterdam, San Francisco, Seoul, Hamburg, Lisbon and Madrid committed to reduce these cities' carbon emissions.
Solar Credits Programme	Subsidy for the instalment of solar photovoltaic systems on Australian houses.
Solar Leasing	Subsidy for the instalment of solar photovoltaic systems on Singapore housing blocks.

Sustainable Backyard Program	Collaboration between the City of Chicago, an NGO and garden material suppliers. It aims to improve the environmental sustainability of homeowners' gardens.
Sustainable Business Leader Program	Boston-based best-of-class benchmarking tool. It aims to stimulate local small and medium-sized businesses to improve their environmental performance and that of their workplaces.
Sustainable Public Procurement	Commitment of governments in the Netherlands to achieve 100 per cent sustainable procurement of their office materials and workspace by 2015.
Transition Towns	Global network that aims to mobilize community action, and foster community engagement and empowerment around climate change.
Urban Resilience Building Codes of New York	Set of building codes that seeks to improve the urban sustainability of New York-based buildings. Introduced after Hurricane Sandy.
Waste Framework Directive	European-wide directive that requires the 27 European Union member states to meet a 70 per cent recycling target for construction and demolition waste by 2020.

In the chapters that follow I predominantly address the countries introduced before, and examples of governance tools discussed with those 500 experts. This large set of countries and the large set of governance tools provide a window on the opportunities and constraints governments and NGOs face in governing urban sustainability and resilience – even though it is not a perfectly representative sample of all the countries in the world, and all the governance tools for urban sustainability and resilience implemented.

Each of the main chapters of this book addresses a particular approach to governance (Chapters 2 to 4) and is introduced by a discussion of the relevant governance literature, which is then illustrated with real-world governance tools that address urban sustainability and resilience. The theoretical structure of these chapters will be of interest to scholars who work on the edge of urban planning and governance and are interested in the opportunities that particular governance tools provide in achieving urban sustainability and resilience. The wide range of governance tools and the way different actors seek to overcome the three main governance problems that stand in the way for meaningful urban sustainability and resilience will be of interest to practitioners and policy makers in this field.

The Appendix provides more insight into the methodology and research design of the study presented in this book.

1.5.4 Outline of the Book

Chapter 2 begins by addressing the most common approach to governing urban sustainability and resilience: direct regulatory interventions. This chapter seeks to understand why governments are relied upon in addressing urban sustainability and resilience. Building on the existing governance literature, it discusses the main governance tools applied by governments: statutory regulation; subsidies; taxes; and other economic incentives.

After this brief literature review it turns to a series of real-world examples of direct regulatory interventions that seek to achieve urban sustainability and resilience. It finds that statutory regulation, as expected, faces problems related to lengthy development times, grandfathering and enforcement. Subsidies and other economic incentives appear to hold a promise to address these problems related to grandfathering to a certain extent, but face their own complications. Subsidies, for example, can be harmful when they do not take into account the financial risk of the subsidizer due to increased extreme weather events. In order to address problems related to the enforcement of statutory regulations many countries around the world have privatized the enforcement of construction codes. This comes, however, with a number of specific shortfalls too. Especially the accountability of private sector building inspectors may be at risk when they themselves face low levels of oversight.

Chapter 3 is interested in collaborations between governments, businesses and civil society groups and individuals. Building on the existing governance literature, it first seeks to understand why and how businesses and civil society groups and individuals have become active in the governing of urban sustainability and resilience; and, what new roles governments have taken up in their collaborations with businesses and civil society groups and individuals.

After reviewing the literature the chapter turns to a series of real-world examples of collaborative governance tools. It recognizes some of the public participation and collective action problems related to collaborative governance that have been identified in earlier literature since the 1960s (Arnstein, 1969; Olson, 1965). For instance, a true sharing of power is often not accomplished in collaborative governance, and collaborations initiated and led by governments in particular may be subject to political swings. Collaborations further run a risk of becoming elite 'networks of pioneers for pioneers' that exclude non-participants from lessons learned (Kern and Alber, 2010). There also appears to be limited insight into the actual outcomes of these collaborations, and though highly inspirational it remains questionable whether collaborative governance is a generally successful approach to increasing urban sustainability and resilience.

In Chapter 4 the focus is on voluntary programmes and market-driven governance. Over the last decades these approaches to governance have become increasingly popular. Following the structure of the two previous chapters, Chapter 4 first presents a brief review of the governance literature on these two related approaches to governance. It then turns to a series of real-world examples of voluntary programmes and market-driven governance tools that seek to achieve urban sustainability and resilience. For heuristic purposes these examples are clustered as: best-of-class benchmarking tools; green leasing; private regulation; innovative financing; contests, challenges and competitive grants; intensive behavioural interventions; and sustainable procurement.

The chapter finds that there is no shortage of governance tools that build on voluntary participation combined with clear financial rewards for their participants. Perhaps one of the most intriguing findings is that governments play important roles in the development and implementation of these tools. It is of further interest that many of these tools resemble the structure of those developed and implemented under more traditional governance approaches. As with the collaborative tools discussed, however, it remains questionable whether voluntary programmes and market-driven governance achieve meaningful results.

Chapter 5 brings the empirical insights from Chapters 2 to 4 together and uncovers the five main trends in contemporary governance for urban sustainability and resilience. It begins with a closer analysis the 68 real-world examples discussed in the preceding chapters, aiming to find trends in contemporary governance for urban sustainability and resilience. It explores:

- Which actors initiated the various tools discussed, and which actors are participating in them?
- What goals do the tools exactly seek to achieve?
- What are the exact roles of the various participating actors?

From this analysis it becomes clear that business and civil society now play a major role in governing urban sustainability and resilience. Yet, governments, and particularly city governments, play key roles in the current tools that have been implemented around the globe, including the innovative tools discussed throughout the book. That is not to say that city governments are involved enough in the governing of urban sustainability and resilience.

City governments may wish to embrace the current collaborations, voluntary programmes and market-driven governance tools even more. As facilitators and supporters they may help businesses and civil society groups and individuals to develop innovative governance tools. More importantly, as assemblers they may seek synergies between different tools, aiming to ensure that the whole of governance tools for urban sustainability and resilience will be larger than the sum of its parts. Then, building on the theory and the wide range of examples discussed throughout the book, the chapter rounds up with discussing 12 design principles for successful governance of urban sustainability and resilience.

Chapter 6 concludes this book by considering the broader significance of the various approaches to governance for urban sustainability and resilience discussed in Chapters 2 to 4. It seeks to (briefly) answer the main question that drives this book. I address the need for a brave application statutory regulation to speed up the retrofitting of existing buildings, which make up over 98 per cent of our cities; a need for more adequate enforcement of traditional and innovative governance tools; the need to embrace and respond to the heterogeneity of the building sector; and, the new roles that governments, businesses and civil society need to take up if a meaningful transition towards urban sustainability and resilience is to be achieved.

2. Direct regulatory interventions

Buildings are likely to have been among the first regulated entities in the world. In ancient Babylonia, King Hammurabi (*c.* 1750 BC) had a clear vision on the construction of buildings. His set of 284 laws, known as the Code Hammurabi, is regarded as one of the oldest preserved sets of direct regulatory interventions in the world. The code sets, among others, rules regarding a builder's duties and responsibilities towards his client: 'If a builder builds a house for some one, and does not construct it properly, and the house which he built falls and kills its owner, then that builder shall be put to death' (King, 2004, p. 21).

It was not until the migration of population into cities that city builders started to think seriously about housing. Cities, such as Amsterdam in the Netherlands and London in the United Kingdom, were evolving rapidly as their economic progress attracted many prospective citizens. In those days, houses in cities were often built in a similar way as houses in the countryside, with timber and straw. Yet, such houses in great numbers caused a major fire risk, that is, a human-made hazard to which such cities were unlikely to be resilient. And indeed, devastating fires such as the 1452 Amsterdam fire and the 1666 Great Fire of London almost fully destroyed these cities. Both fires sparked governments to draw up far-reaching building and zoning regulations.

Directly after the 1452 fire the Amsterdam city government decreed that from then on houses had to be built of brick instead of wood and that thatched roofs had to be replaced by tiled roofs. This was an unprecedented governmental intervention (Breen, 1908). And, directly after the 1666 fire, The Rebuilding of London Act 1667 was passed, which also spells out that buildings from then on had to be built of brick because 'in reguard the building with Bricke is not onely more comely and durable but alsoe more safe against future perills of Fire' (Charles II, 1667 [1819], section V) – which may very well be one of the first references to urban resilience.

Today, governments seek to control many more aspects of buildings, from structural and fire safety to the prevention of disability discrimination and environmental and resource sustainability. They may approach this in different ways. One is through direct regulatory interventions such as making the use of certain polluting building materials unlawful, subsidizing the instalment of solar panels on existing buildings or reducing the land tax for buildings with high levels of environmental performance. It is clear that choosing the right governance tool matters if effective outcomes are to be achieved (Baldwin et al., 2011; Stewart, 2006). But which is the right tool for governments in what setting?

Building on the vast literature in this field and a series of real-world examples, this chapter seeks to better understand the promises and limitations of direct regulatory interventions for achieving urban sustainability and resilience, and since enforcement is such a pivotal part of direct regulatory interventions (but also for the innovative tools discussed in Chapters 3 and 4) particular attention is paid to it in this chapter.

2.1 CHARACTERISTICS OF DIRECT REGULATORY INTERVENTIONS

2.1.1 Actors Involved

It goes without saying that the actors involved in direct regulatory interventions are predominantly governments, governmental departments and governmental agents. But why rely on governments for addressing 'to shape, regulate or attempt to control human behaviour in order to achieve a desired collective end' (as per the definition of governance introduced in Chapter 1), for instance, urban sustainability and resilience?

It is often argued that governments are needed to establish and maintain property rights, and to address market failures (Baldwin et al., 2011; Parkin et al., 2005; Witztum, 2005).[1] Property rights give certainty, which is considered a necessary condition for economic development (Galiani and Schargrodsky, 2010). Property rights may secure land ownership as well as the (future) function of this land, which may give investors or developers enough confidence that taking the risk of developing a building on that land will give them sufficient returns. In terms of urban sustainability and resilience, property rights are of key importance as well. After all, through the issuing of land, zoning regulations and additional regulatory requirements governments have a major say as to where and how lands are to be developed.

In terms of achieving urban sustainability and resilience, a number of market failures stand in the way. Most eminent are the non-excludability of goods and negative externalities.

Non-excludability of goods refers to situations where it is difficult, if not impossible, to exclude people from using the good. City parks may be considered a typical example of a non-excludable good (although more and more parks appear to becoming gated and only accessible to 'selected' users, for example, Blinnikov et al., 2006).

Negative externalities are situations where the price of a good or service does not account for all gains or losses related to it. Environmental degradation, or climate risks more generally, is considered a typical negative externality (Stern, 2006). For a long time (and as is still the case), the negative consequences of our current levels of production and consumption, for instance, greenhouse gas emissions, are not included in the price we pay for our goods and services.[2]

Other relevant market failures to consider in addressing urban sustainability and resilience are imperfect competition, information asymmetries and bounded rationality (Baldwin et al., 2011; Parkin et al., 2005; Witztum, 2005). Imperfect competition may result in monopolies when only one provider, or a very small number of providers, supplies a market with a particular good or service. A good example is the development of the currently popular best-of-class building benchmarking tools (see Chapter 4). By requiring certain construction materials or certain construction processes, such tools limit market access for some players in the sector while rewarding others with a competitive advantage (Beddoes and Booth, 2012; Gifford, 2009).

Information asymmetries refer to situations where one party in a transaction has better information than the other. For cities this is a major concern because so many aspects that make buildings and urban environments sustainable and resilient are hidden behind ceilings, walls, pavements and so on. This information is often costly to obtain (as it requires partial demolition and reconstruction). Besides, given the often highly technological nature of buildings and urban environments, it is questionable whether consumers especially are able to obtain and understand all the information needed to enter into transactions that are efficient to them.

Finally, bounded rationality refers to the idea that in making decisions people are limited by their cognitive capacities, the information they have and the limited time they have to make decisions (Simon, 1945). This appears to be of particular relevance when addressing urban environmental sustainability. People are unable to foresee the far-reaching environmental impact of their behaviour. Strict direct regulatory interventions, which seek to counter the effects of such behaviour, may then face resistance. Alternative strategies that not only seek to counter the effects of this behaviour but also explain the need to counter it may possibly achieve better outcomes (Gsottbauer and Van der Berg, 2011).

In terms of achieving urban sustainability and resilience, a strong role of government may further be needed since governments are able to apply a very long-term vision, which citizens and businesses often cannot (Deflem, 2008). Through governments, it may further be argued, relatively powerless groups may see their interests served (Skocpol, 1985). Another viewpoint is rather simple, but interesting. It may be argued that societies are fully based on this role of governments, and over the centuries people have become so used to this role of governments that they do not question it anymore. A change of this institutional structure is almost unimaginable. The most logical way forward, then, is to continue giving governments rather far-reaching powers to achieve desired collective ends, which only reinforces the 'systems' we are currently in (for example, Croley, 2011; Pierson, 2004).

2.1.2 Governance Tools

Direct regulatory interventions come in various forms. In terms of governing urban sustainability and resilience, it is of interest to pay attention to statutory regulation, direct subsidies and (other) economic instruments.

Statutory regulation

The classic direct regulatory governance tool, and perhaps most used in governing buildings and urban environments, are statutory regulations. An example is building regulations that set requirements to the structural safety of a building. Such requirements are often expressed in standards that seek to steer behaviour in such a way that harmful results are prevented or that a specific outcome is achieved.

At least three types of standards may be distinguished (Baldwin et al., 2011): prescriptive standards, performance-based standards, and target or goal-oriented standards.

Prescriptive standards seek to prevent harmful events, for instance, the collapsing of a building, by stating the exact requirements for the particular parts of a building, its construction or even its design process have to meet. Typically prescriptive building regulations set standards to the loadings for buildings, for instance, how much load a floor should be able to hold, to the structural use of concrete or other building materials, or to the process of calculating its structural safety. An example is: 'no floor enclosed by structural walls on all sides exceeds 70 m^2' (HM Government, 2010, p. 18).

Such standards require governments to possess far-reaching technical expertise to formulate them. In building regulations this particular issue is, however, often overcome by referring to (international) standards that are developed by non-governmental and private sector organisations. For instance, the German Federal Building Code relies on many standards developed by the non-governmental German Institute for Standardization (Deutsches Institut für Normung, DIN), while building regulations in many other countries refer to standards developed by the International Organization for Standardization (ISO).

Besides, prescriptive standards leave little room for flexibility or innovation, which is an oft-heard critique (Duncan, 2005; Gann et al., 1998). One side effect of this situation is that circumstances may arise where a novel technology, which would enhance environmental sustainability, cannot be applied because the prescriptive regulations do not allow for it – recall the example of grey water use introduced in Chapter 1 (Section 1.4).

Performance-based standards partly overcome this critique. These standards specify the performance of a good or service, but do not specify how that performance is to be achieved. Such standards are normally considered to give those regulated an incentive to find a solution that is both effective in terms of meeting the standard and efficient in terms of costs (May, 2003; Meacham et al., 2005).

In the Netherlands, for example, the Dutch Building Decree sets a design target for the energy efficiency of (future) houses, but does not stipulate how this target is to be achieved (Beerepoot and Beerepoot, 2007). Since its introduction in 1995, the stringency of this target has been increased and it is to be expected that houses built from 2020 onward have to meet a zero-energy target. Interestingly, however, recent studies show that the actual energy use in Dutch houses, built since 1995, is different from what the regulatory standards require, and household behaviour is considered a major reason for the mismatch between regulated performance and actual performance (Guerra Santin and Itart, 2012).

Finally, target or goal-oriented standards seek to prevent harmful events by directly linking the behaviour of individuals, goods or services to the regulatory goal. They leave it fully to those regulated as how to achieve compliance with these standards. The Building Codes of Australia, for instance, express the objective of fire safety as:

> To (a) safeguard the occupants from illness or injury (i) by alerting them of a fire in the building so that they may safely evacuate; and (ii) caused by fire from heating appliances installed within the building; and (iii) in alpine areas,

> from an emergency while evacuating the building; and (b) avoid the spread of fire; and (c) protect a building from the effects of a bushfire. (ABCB, 2009, p. 3401)

The Building Codes of Australia, however, do spell out various approaches to meet this goal-oriented standard.

It goes without saying that different types of standards can be combined. In Singapore building energy codes specify requirements for, among others, the energy efficiency of the building envelope, installations and water efficiency. Points are awarded for meeting particular requirements and a certain number of points are required for a building permit. In order to stimulate sustainable building design and development, bonus points are awarded for the use of renewable resources (BCA, 2008).

This is an innovative play on combining prescriptive and goal-oriented standards. In India the Bureau of Energy Efficiency is experimenting with another innovative approach to building regulation. It has made a separation between the energy performance of the construction (that is, building structure, walls, roofs, windows and the like), the energy performance of large building installations, such as heating, ventilation and air conditioning (HVAC) systems, and the energy performance of smaller installations that are more difficult to capture in building regulation, such as lightning. Different authorities may be involved in the enforcement process of these different 'bundles' of regulatory requirements (Rawal et al., 2012).

Direct subsidies

Direct subsidies are often a form of financial support aiming to promote beneficial economic or social outcomes (Myers and Kent, 2001). Subsidies may help to keep consumers' prices for a good or service below market level or keep producers' prices above market levels. Subsidies thus provide a different incentive than statutory regulation.

The latter seeks to prevent harmful events by making those regulated fear the consequences of non-compliance, or by setting or changing the (societal or internal) norm as to what is acknowledged as accepted behaviour (Hawkins, 1984; Kagan, 1984; Tyler, 1990). Subsidies seek to prevent harmful events by financially supporting those regulated to avoid these events occurring.

This may be done directly by subsidizing housing for low-income households. In this example the subsidies are related directly to the regulatory goal: ensuring that low-income households are able to afford housing. It may also be done indirectly by subsidizing the instalment of solar panels by households. In this example the subsidies may serve different goals, and may do so indirectly: support the market for solar panels by increasing household demand; address the negative consequences of using fossil fuels by supporting a transition to renewable energy; and, change households' attitude towards solar panels by making them a more normal aspect of daily life when more and more people install solar panels on their houses.

Although subsidies have a long history of application, they are controversial (Baldwin et al., 2011; Myers and Kent, 2001; Pearce et al., 2003). In terms of implementation, subsidies may be critiqued for a potential danger of unequal distribution. After all, subsidies only give an advantage to those who are able to obtain and understand the information related to such subsidies.

There is a risk that subsidies will not achieve their goals: what can governments do if the money is spent but the regulatory outcome not achieved? (Baldwin et al., 2011). Subsidies are sometimes even considered harmful. A typical example is the subsidizing of fossil fuels. As a result of direct subsidies to producers and consumers, the price of fossil fuel does not reflect its actual production costs, let alone the costs of environmental harms caused by using fossil fuels (UNEP, 2008a). With the built environment taking up such a large share of energy consumption (about 40 per cent of all energy used globally; see Pérez-Lombard et al., 2009), these 'harmful subsidies' may significantly distort the implementation of more sustainable and resilient energy sources in urban settings.

Economic instruments

Besides direct subsidies, governments apply a range of other economic instruments to steer behaviour. The two most well known are taxes and tradable permits.

In terms of achieving environmental sustainability, governments around the globe have been implementing environmental taxes (Stewart, 2006). Such taxes seek to correct the price of production and consumption by including the costs of negative externalities. For instance, in a number of European countries taxes apply to the extraction of sand, gravel and rock for the cement industry. The environmental costs of these activities would normally not be included in the price of cement and the taxes seek to address this particular issue (EEA, 2008).

In applying such taxes, governments may seek to make harmful practice less profitable or to use the revenue to restore the harm done. This, then, is also one of the major criticisms of environmental taxes: they may create the illusion that harmful behaviour is accepted because the behaviour is paid for or they may create an industry culture where the tax is being considered as just one of the many costs of doing business (Harper, 2007; Zarsky, 1997).

In line with environmental taxes, tradable permits seek to overcome market failures. However, they do not only correct the price of production and consumption but also often seek to put a limit to the amount of negative externalities. Carbon emission trading is a typical example. Under a carbon emission trading scheme a government may set a maximum (a 'cap') to the carbon emissions it expects to be produced. It can then issue permits that allow the holder to produce a certain amount of carbon emissions. If this holder produces less than it is allowed to, it can trade its permit with a producer that seeks a larger quota of carbon emissions than it holds under its own permits. It is then expected that a price will be achieved that expresses the market costs of carbon emissions. Further, under a tradable permit scheme it is expected that producers will seek modes of production that (cost-)effectively reduce their carbon emissions below the cost of buying permits, for instance, by occupying energy-efficient buildings (for extensive discussions, see, among others, Aakre and Hovi, 2010; Ishikawa et al., 2012; Pope and Owen, 2009).

2.1.3 Development, Implementation and Enforcement

The development and implementation process of direct regulatory interventions is, in theory, rather straightforward. Governments develop direct regulatory interventions, monitor compliance with these and take disciplinary measures if non-compliance is observed. However, it almost goes without saying that reality is more complicated, and the development and implementation process of direct regulatory interventions varies from country to country (Baldwin et al., 2011; Wilson, 1989).

In some countries regulators involved in the implementation of direct regulation for urban environments also hold significant development powers, such as in the United States; while in other countries the process of development and implementation are more separated, such as in the United Kingdom. There is something to say for both approaches. If regulators are involved in both implementation and development of direct regulatory interventions, they may include the lessons learned from implementation in the development process of (future) tools (Mytelka and Smith, 2002). This feedback loop may get lost if different governmental actors or agencies are involved in the development and implementation processes. However, the risk of keeping these processes in the hands of the same actors or agents may result in situations where decision makers build close relationships with those they seek to govern, and are, consequently, influenced by the latter. Such 'regulatory capture' is repeatedly discussed in the regulatory literature (for overviews, see Baldwin et al., 2011; Levi-Faur, 2011).

Regulatory capture illustrates one of the key paradoxes of governing through direct regulatory interventions. Those governing often need input from those governed to be able to develop effective and efficient governance approaches. After all, those governed will likely hold more information on their practice and behaviour than those governing. Yet, it may not be in their interest to share this information with those governing, and by giving partial or misleading information they may seek to steer those governing in such a way that their own interests are served, perhaps over the best public interest.

It is not surprising that this paradox is sometimes discussed in terms of being 'a game between inspectors [that is, governors] and inspectees [that is, governed]' (De Bruijn et al., 2007). Interestingly, as further addressed in Chapter 3, this close interaction between governing and governed actors is considered key to achieving successful governance outcomes under a collaborative governance approach.

Enforcement

Throughout this book it will become clear that without adequate enforcement of traditional and innovative governance tools for urban sustainability and resilience not much should be expected of their outcomes. Over the years a large literature has developed that exactly addresses the enforcement of statutory regulation and other approaches to governance.

In relation to the enforcement of statutory (building) regulation in particular the work by John Braithwaite (further discussed in, Ayres and Braithwaite, 1992) and Neil Gunningham (further discussed in, Gunningham and Grabosky, 1998) is of interest.

In the book Responsive Regulation (Ayres and Braithwaite, 1992), Braithwaite argues for a staged approach to enforcement in which persuasion and punishment are combined. This approach starts with persuasion (i.e., dialogue, educational strategies, guidelines and protocols) and ends with the most severe punitive measure available (i.e., revocation of a licence to work, imprisonment, high financial penalties). Braithwaite (together with Ian Ayres) considers that punishment as first choice is unaffordable, unworkable and counterproductive. Governments generally have limited enforcement capacity, and not all who are subject to regulation need a similar level of enforcement to ensure compliance. Yet, fully rejecting punitive incentives is considered naïve as well. This had become clear after years of experimenting with self-regulation in which those subject to statutory regulation monitored their own compliance.

The trick in regulatory enforcement, according to the responsive regulation philosophy, is to combine the stick and the carrot. Instead of applying a one size fits all enforcement strategy it is of more avail to question: Where can those subject to governance tools be supported in achieving compliance? How can they be pursued in complying with regulation? What is the ultimate enforcement measure that induces enough fear, even when regulators do not have to use it all too often?

In the book *Smart Regulation* (Gunningham and Grabosky, 1998) Gunningham builds on Braithwaite's work and considers that governments are not always in the best position to carry out regulatory enforcement. In this book Gunningham (together with Peter Grabosky) divide the regulatory process into parties, roles and interactions. The key to the smart regulation philosophy is to have those actors involved in the regulatory process that are best fit to enforce statutory regulation, and governance tools more generally. Sometimes this may be through traditional government agencies; sometimes through self-regulatory or co-regulatory initiatives in which private sector actors enforce their own body; sometimes through third parties, such as consumer interest groups or insurance companies that act as 'surrogate regulators'.

From these two works it is a little step to what have become known as networked enforcement regimes (Baldwin and Black, 2008; Braithwaite, 2004; Drahos, 2004, 2013; Grabosky, 2013). The philosophy underlying such regimes is to have the various stages from persuasion to punishment from the responsive regulation strategy carried out by those actors who are best suited to do so. Key is for government to find these actors and include them in the development of direct regulatory interventions (and other governance tools) and in their enforcement practice. Readers interested to see what such a regime may look like in the building sector may wish to go to the concluding pages of this book where I present a diagrammatic illustration.

2.2 FROM THEORY TO PRACTICE: URBAN EXPERIENCES WITH DIRECT REGULATORY INTERVENTIONS

This section addresses experiences with statutory regulation and other direct regulatory interventions on international, national and local levels to gain a better understanding of how direct regulatory interventions play out in real-world settings, and what the roles of governments are in these.

2.2.1 Statutory Regulation

Twenty-seven European countries, one set of energy regulations for buildings

Battling climate change, with the targets of the Kyoto Protocol in mind, the European Commission has introduced and implemented a range of policies and programmes to improve the environmental performance of its building sector and its built environment (for an overview, see WGBC, 2011).

The most far-reaching attempts to do so are a range of directives aimed at harmonizing the building regulatory frameworks in European Union member states. Such harmonization serves a dual goal: on the one hand, it decreases current barriers to the free trade of goods and services among European Union member states (an economic goal); on the other hand, it provides the European Commission the opportunity to address societal risks such as climate change on a European level (a social goal).

The best-known European Commission attempt to harmonize its member states' sustainable building regulatory frameworks is the Energy Performance of Buildings Directive (EPBD). This directive was issued in 2002 and recast in 2010. The original directive requires, among other things, that member states set minimum energy performance requirements for residential and commercial buildings – for new construction work and for major renovations. Further, the 2010 recast requires that member states ensure 'nearly zero energy buildings' by the end of 2020; provide fiscal and financial incentives to encourage sustainable construction that complies with higher energy levels than regulated; and requires energy performance certificates to be provided in all buildings and be displayed to the public (EC, 2010).

Other illustrative directives are the Energy End-use Efficiency and Energy Services Directive, which requires member states to draw up national action plans to achieve 1 per cent annual energy savings over nine years starting in 2008 (EC, 2006), and the Waste Framework Directive, which obliges member states to meet a 70 per cent recycling target for construction and demolition waste by 2020 (EC, 2008). It is through the transposition of these directives in the member states' national building regulatory regimes that they come into effect.

In 2010 an evaluation of the EPBD was carried out to better understand how the European Union member states have implemented this directive into their building regulatory regimes (Van der Heijden and Van Bueren, 2013; Vermande and Van der Heijden, 2011) – 24 member states participated in this study. It was found that all these member states have introduced statutory regulations related to the energy performance of buildings.

Delving a bit deeper into the study, however, it becomes clear that the effectiveness of these direct regulatory interventions is questionable. Member states seem to leave it largely to the building sector to determine the best way to meet the new energy performance requirements. Although this fully fits the concept of goal-oriented regulation, some guidance may help these industry players in meeting these goals. When such guidance is lacking it will be unclear for those governed to know when they comply with statutory regulations and when not – an issue that was found particularly relevant in a rather new area, energy performance and reduced energy consumption. Also, enforcement of these regulations appears to fall short in most of the member states that participated in the study. Although respondents referred to some enforcement of energy regulations during the assessment of building plans, most of them highlighted an absence of enforcement of energy regulations during the construction phase of a building. It may very well be that already busy building authorities do not have the manpower to take up the enforcement of additional regulatory requirements (see also Section 2.4).

The EPBD only addresses new buildings and major renovations. It does not apply to the existing building stock. In 2013 the European Commission once more stressed that if the European Union is to meet its targets of carbon emission reductions it needs to address the full building stock (EC, 2013). In terms of addressing existing buildings, it appears to step away from mandatory retrofits (but see some first ideas about a 'renovation roadmap for buildings' in Klinckenberg et al., 2013) and focuses more on innovative financial tools such as soft loans and competitive grants. As the European Commission states:

> building owners will have to be convinced of the benefits of making their properties more energy efficient, not only in terms of a lower energy bill but also as regards improved comfort and increased property value. This may well be one of the most important hurdles to overcome in making Europe's buildings more energy efficient. (EC, 2013, p. 12)

Mitigating the risks of extreme weather in Australia

As in many countries, extreme weather events are becoming more frequent and more severe in Australia (Cork, 2010; Senate Standing Committees on Environment and Communications, 2013). An example is a series of floods that hit the state of Queensland in December 2010 and January 2011. With at least 70 towns flooded, over 200 000 people affected and an estimated damage of A$2.38 billion, it is considered Australia's most devastating flooding in terms of damage to urban environments (Carbone and Hanson, 2013).

Australian governments acknowledge the risks of such weather events and work towards policies that seek to reduce risks (COAG, 2009; Productivity Commission, 2012; Queensland Reconstruction Authority, 2012; Senate Standing Committees on Environment and Communications, 2013). The Australian National Construction Code of 2012 (in force since May 2013), for example, sets requirements that apply in flood hazard areas. As is often the case with construction codes, these requirements are drawn up as statutory regulation. For instance, to prevent buildings being critically affected by floating debris during a flood, it requires that 'structures must be determined using engineering principles as concentrated loads acting horizontally at the most critical location at or below the DFL [defined flood level]' (ABCB, 2012, p. 15).

The current code also lays down requirements for buildings to resist wind, snow, earthquakes, rainwater pressure, landslides and bushfires. This broad scope is praised by the United Nations since it 'tick[s] the boxes' in addressing relevant disaster resilience elements (UN, 2013b, p. 3). Yet, not all actors in the Australian building sector appear satisfied with the code (Deloitte, 2013).

The requirements are considered as insufficient to mitigate the risks posed by recent natural hazards in Australia such as bushfires and storms (Australian Resilience Taskforce, 2012). A reason for this mismatch may be found in the origins of the code: it aims to represent minimum standards and changes to the code have to be approved by the Office of Best Practice Regulation, an agency within the Australian Government Department of Finance (Yates and Bergin, 2009). This may result in a shift in focus when proposed regulations move from one agency to the next – from seeking maximum safety to seeking minimal regulatory burden and costs. As a result, so claims the Insurance Council of Australia:

> [the code] permits the construction of buildings (at a minimum standard) that include no element of durability (property protection), creating a stock of buildings that while 'safe' [in terms of structural safety as specified in the Building Codes of Australia] are increasingly brittle to extreme weather events. (Whelan, 2012, p. 5)

The Australian building sector is also critical of the government's lack of attention to the existing building stock. It considers that

> [w]hile planning reform and enhanced building codes are an important element of building resilience, they only affect new and renovated homes. The greatest impact of resilience measures, and arguably the biggest coordination challenge, lies with existing residential buildings (retrofit, compliance and relocation). It is often technically difficult and costly to retro-fit an existing property to be disaster resilient. Although one of the hardest to implement, this is also one of the most important areas for resilience action. (Deloitte, 2013, p. 134, 27).

This critique is all the more relevant considering that about 80 per cent of Australia's housing stock does not (have to) meet the statutory building regulations that have been introduced since the 1980s (Yates and Bergin, 2009). It is not expected that households feel an incentive to voluntarily bring their homes into compliance with resilience requirements as expressed in current day building regulations, and more is expected from reforming property insurance (Productivity Commission, 2012). For instance, by mandating a particular level of insurance against natural hazards, households are given the choice to pay a significant premium to insure their 'dangerous' homes, or to retrofit the house and face a lower premium because of its lower risk to insure.

Making New York more resilient

Hurricane Sandy has been a rude awakening to New York. The city is not unfamiliar with coastal storms, but

> with 43 deaths; 6,500 patients evacuated from hospitals and nursing homes; nearly 90,000 buildings in the inundation zone; 1.1 million New York City children unable to attend school for a week; close to 2 million people without power; 11 million travellers affected daily; $19 billion in damage … Sandy was an unprecedented event for New York City. (City of New York, 2013, p. 11)

In the wake of Sandy, New York's former Mayor Bloomberg asked the non-governmental Urban Green Council to convene the Building Resilience Task Force (the Urban Green Council is further discussed in Chapter 4). This task force was asked to study how to improve city-wide infrastructure and building resiliency, as well as how to help communities become more resilient. The task force brought together over 200 individuals from various governmental and non-governmental organizations, as well as businesses. Interestingly, it advised then-Mayor Bloomberg to make a number of significant changes to the building codes of New York (Urban Green, 2013).

In response to the task force, Bloomberg passed a number of proposed changes to the building codes. Among others, the codes now mandate that: fire protection equipment is located above the potential flood level; that equipment and structures added to existing buildings meet the same standards as those for new buildings; and that new toilets and faucets in buildings are able to operate without an external supply of electrical power. The task force notes that these requirements would best also apply to existing buildings, but understands the difficulty in making existing buildings subject to new statutory regulation (Urban Green, 2013).

Yet, even in a city as progressive as New York it is unlikely that retrofits to existing buildings will soon become mandatory. A 2009 proposal by Bloomberg to require mandatory energy retrofits for existing buildings was opposed as being too expensive for property owners in a time of global economic stress (Cheatham, 2009). Instead, a series of other regulations were introduced. Under the Building Energy Code, renovations have to meet minimum energy conservation standards; energy and water use benchmarking is compulsory in certain city-owned and privately owned buildings (depending on their size); and energy audits and retro-commissioning are mandatory for all existing buildings that exceed 50 000 square feet (City of New York, 2009).

Addressing slums and illegal urbanization in India

In 2013 about 900 million people lived in slums, an astonishing 12 per cent of the world's population (UN-HABITAT, 2013).[3] The people living in slums have often moved from the countryside to a city hoping to find a better future. They are unlikely to have sufficient funds to build or rent a house that meets high (and costly) construction standards (Murthy, 2010; Sen, 2013). Slums are particularly vulnerable to natural and human-made hazards. Slum buildings are often built illegally, not following (statutory) construction codes or planning and zoning regulations. When disaster strikes, it is very likely to strike hard in slum areas (IFRC, 2013; UN, 2013a).

A typical example is the collapse of a building in the Mumbra area of Mumbai, India, on 4 April 2013; 78 people died in the incident, including 18 children. The seven-storey building was built illegally on land not designated for building development. It was built quickly with cheap and poor quality building material and housed allegedly 100 to 150 people (Prakash, 2013; Udas, 2013) and, as is often the case, the builder bribed officials to not take action against the building. The building certainly was not the only illegal building in the precinct: 'I have been told 70 per cent of buildings in these areas are illegal', a government official noted days after the incident (quoted in NDTV, 2013). The Thane district, of which the Mumbra area is a part, houses approximately 500 000 illegal buildings (Thomas and Yeshwantrao, 2013). With such large numbers of illegal buildings governments simply cannot take the most drastic action. Demolishing these illegal buildings implies that thousands of people have to be made homeless. The logical step these people take is to build or rent illegal shelter in the next slum area.

Governments facing slum problems find themselves in a difficult position. They need housing provision for their rapidly growing urban population. This housing needs to be, at the very least, structurally sound. Existing slums do provide housing, but the housing quality is poor. However, governments cannot enforce the construction codes on buildings that have been built in violation with planning and zoning regulation. In addition, some recent research indicates that those owning buildings do care more about the durability of their property when there is no threat of potential demolishment (Minnery et al., 2013) – that is, why would they invest in a durable but costly building if the government has the right to demolish it and threatens that it will do so?

International organizations such as the United Nations and the Red Cross call for a change of property rights in slum areas (IFRC, 2013; UN, 2013a). By giving slum dwellers property rights, or by at least legalizing slums as urban areas, governments may provide an incentive for the development of more durable buildings. Legalizing slums does nevertheless come with its own 'grandfathering' problem: structurally unsound houses and built-up areas remain unsound until further action is taken. Legalizing a slum may create the illusion that the government has dealt effectively with an illegal and dangerous situation while in fact it has only changed the administrative status of the situation. Besides, legalizing slums may incentivize people to start new slums, hoping that these will be legalized in the future, and it gives governments formal responsibility over public areas, public health and public safety in physically complicated urban settings (Ghadyalpatil, 2006; The Hindu, 2005).[4]

2.2.2 Subsidies

Subsidized solar power in Germany

Germany is often lauded for its progressive energy regulation (Bechberger and Reiche, 2004; Frondel et al., 2010; Lipp, 2007). In particular, the 1990 'Feed-in-Law' has had a significant impact on the built environment. This law guarantees fixed prices for energy generated from renewable resources for 20 years, and mandates grid operators to connect renewable energy installations to the net and feed in all generated energy and pay the specified price (Reiche, 2005). This long-term stable price provides market security. The high feed-in tariff in Germany has resulted in a large uptake of solar photovoltaic systems by German households (Dohmen et al., 2013; UPI, 2013; Wynn, 2013).

The Feed-in-Law, and in particular the subsidizing of solar power, has however come under criticism (Frondel et al., 2008). The total electricity generated by solar power in Germany was about 0.6 per cent in 2008, whereas that of wind power was about 6.3 per cent in the same year (Frondel et al., 2010). This means that solar power only marginally adds to reduced non-renewable energy consumption and carbon emissions in Germany.

The subsidies to support renewable energies are favourable for solar power: the feed-in tariff of solar power is about six times that of wind power, and about ten times higher than the market price of conventionally produced energy (Lipp, 2007). In short, installing solar photovoltaic systems is highly profitable for those owning the system, but very expensive for German taxpayers, whose tax money potentially funds a form of renewable energy that is not the most cost-efficient yet. Roughly 25 per cent of all feed-in tariff expenditures are allocated to solar power, which makes up less than 10 per cent of all renewable energy produced (Frondel et al., 2010).

It is expected that it will be at least 2020 before the cost of energy produced by solar power in Germany will be on a par with that of conventional energy production; a result of photovoltaic technology becoming more efficient and effective, with conventional energy production becoming more expensive (Hoffmann, 2006). Paradoxically, it seems that highly subsidized solar power in Germany has created a worldwide market for solar power technology, and that the high German demand for photovoltaics has resulted in significant global reduction of the cost of solar power technology (Wright, 2013).

Rooftop cowboys in Australia

Where Germany may not seem to be the most logical country to push for solar power technology, with its relatively low insolation (exposure to the sun), few will argue about Australia's potential for utilizing solar power in the built environment. Through the Solar Credits Programme the Australian Government has sought to increase the uptake of solar technology by households.

The programme subsidized the instalment of solar photovoltaic systems by about 50 per cent (combined data from Australian Government, 2012; Martin, 2012). In combination with favourable feed-in tariffs, offered in various Australian states (Energy Matters, 2013), the programme became so successful that it had to be terminated six months early, in November 2012, because of excessive demand for subsidies

(ABC News, 2012). In 2013 various Australian states cut back the original generous feed-in tariffs since the original tariffs put state governments at financial risks (Lloyd, 2013). In addition, much critique was raised as to the fairness of the programme (McIntosh and Wilkinson, 2010). The high feed-in tariffs were recovered by raising the price of conventional electricity supply, which meant that households without solar photovoltaic systems were financially supporting those who had solar photovoltaic systems installed.

The programme was also plagued by a series of fraud cases. In September 2011 the Australian Competition and Consumer Commission issued a warning to consumers to be aware of scam solar offers (ACCC, 2011). It mentioned fraudulent acts, such as scammers offering free or highly subsidized solar photovoltaic systems, asking for an upfront fee, but not delivering on their promises. Other sources mention the instalment of low-quality solar photovoltaic systems, or simply poor instalment work (Guest, 2011; Madigan, 2010).

The technical complexity and unfamiliarity of Australian households with solar photovoltaic systems, combined with the complexity and generosity of the subsidies and feed-in tariffs, prompted the then Australian Fair Trades Minister, Antony Roberts, to state that the solar industry had become a 'total magnet for cowboys' (quoted in Smith, 2011). Stories on 'rooftop cowboys in the solar industry' with fraudulent contractors misleading homeowners about the costs of the instalment of solar panels under a subsidy programme have attracted much media attention in Australia (Peacock, 2013).

Harmful subsidies in the United States

Hurricane Sandy has also been a rude awakening to many coastal homeowners, some of these in New York, partly in the form of changes to the National Flood Insurance Program that came in force on 1 October 2013 – a year after Sandy.

This programme was established in 1968 and requires mandatory flood insurance of homeowners who have federally backed mortgages (Chivers and Flores, 2002). The programme was developed because private insurers were unwilling to take the market risk of providing flood insurance. Homeowners simply were not willing to pay the premium for this insurance at market rates. To make sure that homeowners would be sufficiently insured, the federal government introduced the programme, which offers subsidized insurance to homeowners (Dixon et al., 2013). However, with an increasing number of homes being built in coastal areas since the 1970s, the financial risk of this programme became substantive. Before Sandy hit the United States the programme was already US$17 billion in debt resulting from payouts due to the series of hurricanes that plagued the United States in 2005 (Katrina, Rita and Wilma), and it is expected that this debt will grow to an astonishing US$30 billion due to payouts as a result of Hurricane Sandy (Union of Concerned Scientists, 2013).

To recover from these debts, and to mitigate future financial risks, the programme's rate will go up 25 per cent a year until it reaches levels that actually reflect the risk from flooding. Instead of paying about US$500 a year for flood insurance, homeowners are more likely to face fees that reach into thousands of dollars yearly. This change to the programme has caused severe civil unrest (Walsh, 2013).

This example highlights, once more, the potential harmful effects of subsidies. Would households have chosen to live in flood-prone areas in such large numbers if they were to take the full (financial) risk for doing so?

2.2.3 Economic Instruments

Smart taxes in the United States

Tax-based incentives are another approach to stimulate urban sustainability and resilience (Bakker, 2009). The Canadian Renewable and Conservation Expense incentive (Natural Resources Canada, 2010) provides a 100 per cent income tax deduction for expenses such as studies of suitable sites for building projects that use renewable energy sources, the costs related to getting the relevant permits for developing such projects or the cost of solar photovoltaic systems with a capacity of 3 kilowatts or larger.

The US state of Maryland ran the Green Building Tax Credit Program from 2009 to 2012. The programme provided developers with tax credits for the construction and retrofitting of energy-efficient buildings. The programme was closely linked to Leadership in Energy and Environmental Design (LEED), which is a building benchmarking tool (see Chapter 4). Tax credits would only be issued if a building, upon completion, met LEED Gold requirements, and a LEED accredited professional was to assess the construction work once finished. The programme was terminated in 2012 due to harsh economic conditions, but is said to have lowered energy demand, and reduced energy usage by 30 per cent in a number of projects across the Maryland commercial and residential industries (Maryland Energy Administration, 2012).

An alternative tax incentive is provided in the US city of Cincinnati, in Ohio. Here the city offers property tax credits to homeowners when they improve the energy efficiency of their homes or seek to build a new home that meets particular energy standards (Hirokawa, 2009).

Privileges for developers and property owners in the United States, Australia and India

Another way for governments to directly influence the environmental sustainability of buildings is by privileging sustainable buildings over non-sustainable buildings in administrative processes (Abaire, 2008).

The county of San Diego, in California, runs the Green Building Incentive Program. The programme offers incentives of reduced plan check turnaround time and a 7.5 per cent reduction in plan check and building permit fees for projects meeting programme requirements. These requirements include: reduced material consumption by using straw bales as a construction material; reduced water consumption by using grey water systems; or, energy reduction by significantly exceeding the minimum energy standards as stipulated by the California Energy Commission.

North Carolina, in the United States, allows all its counties and cities to provide reduced building permit fees if buildings meet guidelines established by LEED or another nationally recognized programme. A few jurisdictions in North Carolina are further allowed to provide density bonuses to builders who build or retrofit energy-efficient buildings. Again a link with LEED is made as the standard for assessment (North Carolina General Assembly, 2008).

A density bonus provides a clear economic incentive: the builder is allowed to develop more buildings, or more units on the plot of land than zoning law would normally permit. Density bonuses apply in Arlington County, Virginia. Here the county board can allow an increase in the density of development as well as an increase in the height of development if that development meets a specified LEED rating (Arlington Economic Development, n.d.). Similar incentives apply in the United States in states such as Florida, Massachusetts and Georgia (Freilich et al., 2010).

These incentives are also popular outside the United States. For instance, in Queensland, Australia, under the Green Door programme, construction projects that are identified as 'the most sustainable in Queensland' are fast-tracked in order to ensure 'exemplary sustainable developments delivered sooner throughout Queensland' (Queensland Government, 2011, p. 4). In India a range of incentives is in place for developers and property owners who commit to the voluntary Green Rating for Integrated Habitat Assessment (GRIHA) best-of-class benchmarking tool (see Chapter 4). In some cities GRIHA (pre-)certified projects are fast-tracked through environmental clearance assessments, while in other cities these projects face reduced costs for mandatory plan assessments (GRIHA, 2014).

Providing information on energy performance in Europe

The earlier discussed EPDB mandates the introduction of Energy Performance Certificates across the European Union, among other governance tools that aim for improved energy use in the European building stock. The certificate gives insight into the energy performance of a building based on its thermal performance. It also contains some specific advice on how to improve the thermal performance of the building. The certificate is compulsory at the sale or the start or renewal of a lease of a building that was built before 1999, unless the buyer or (future) tenant of that building signs a waiver (EC, 2010). Particular buildings, such as monuments, are further exempted from this direct regulatory intervention. The certificate is expected to provide transparency on a building's energy performance to consumers. For instance, a prospective homeowner may base her decision of buying a home on the building's energy performance.

The Netherlands was one of the first European Union member states to introduce energy performance certificates in 2008. During the implementation it faced a number of considerable drawbacks that affected the take-up of the certificates (Brounen and Kok, 2011). Just before the launch of the certificates in 2008 the Dutch Association of Homeowners started a media campaign that highlighted the lack of reliability and consistency of the certification process. The media picked up on this campaign and reported negatively on the certificates. Policy makers in their turn did not communicate clearly the advantages of the energy certificate. The escape clause of waiving certification by home buyers added a further complication, and may explain why by the end of 2009 less than 20 per cent of houses in the Netherlands were certified at the time of transaction (Brounen and Kok, 2011).

Seeking to strengthen the uptake of the energy performance certificate in the Netherlands, the responsible minister, in 2011, proposed to mandate the certificates for new buildings. This would give owners of new buildings insight as to whether the energy performance as promised in the building contract was realized in their building. In addition, the minister proposed scrapping the possibility for buyers to waive the energy performance label. Both proposals were rejected by the parliament, which considered that these proposals would result in too much financial burden for the building sector and to undesired additional regulation and bureaucracy (Tweede Kamer der Staten Generaal, 2012).

These negative experiences with energy performance certification are reported in other European member states as well (Casals, 2006; Dixon et al., 2008; Janssen, 2005; McGillian et al., 2008; Mlenik et al., 2010).

A German study (Amecke, 2012) confirmed that the existence of the escape clause negatively affects the uptake of the certificate in Germany. It also mentions that German households, although aware of the certificates, do not seem to see too much value in these as they lack information on the monetary savings of the upgrades advised on the certificate. Such findings are further confirmed by a Belgian–Danish study that finds that homeowners do normally not follow up on the advised upgrades (Gram-Hanssen et al., 2007). A UK-based study, finally, found that enforcement agencies needed time to adapt to the new regulations, which resulted in inconsistent enforcement approaches (see also Ekins and Lees, 2008).

2.3 ENFORCEMENT

2.3.1 Failing Enforcement, Failing Resilience

On 23 April 2013 cracks were discovered in the Ranza Plaza building in Dhaka, the capital of Bangladesh. This eight-storey building was largely used as a garment factory where thousands of people worked. The building was temporarily closed, but the very same day the building's owner claimed: 'This building will stand a hundred years' (Sohel Rana, quoted in Ahmed and Lakhani, 2013).

The very next day the building was in use again; and unfortunately, this day the building collapsed. More than 1100 people died, making it one of the worst structural failures of an individual building in history. Various newspapers reported that illegal construction activities had been going on before the building collapsed, that the building was designed as a shopping mall and not as a factory, and that authorities had fallen short in enforcing building regulations (Ahmed and Lakhani, 2013; Hossain and Alam, 2013; Sattar, 2013). From these newspapers a combination of causes may be found to have resulted in the ultimate collapse: three floors were illegally added to the eight-storey building; day-in day-out the vibrating of the garment machines had put stress on the building's structure; and, finally, 15 minutes before the collapse all of these machines were switched on simultaneously after a brief power blackout, causing extreme vibrations throughout the building.

The collapse of the building has caused, again, insight into the poor working conditions of workers in (rapidly) developing economies. Western clothes brands and retailers have since put much pressure on the Bangladesh garment industry to improve these (BBC News, 2013; Butler, 2013). It has also, once more, provided insight into the poor enforcement practice of construction safety codes.

Five months before the collapse of the Ranza Plaza building in Dhaka, another garment factory had collapsed in Bangladesh, killing 112. The government had promised more stringent inspections of garment factories for structural safety and to pull the licences of factories that would not comply with the regulations. However, the plan for doing so was not implemented by the time the Ranza Plaza building collapsed (Hossain and Alam, 2013). The poor enforcement practice appears a well-known problem in Bangladesh, with one of the factory workers explaining:

> Most buildings don't follow the building code of conduct, so it's easy to get away with these unregulated ways of constructing buildings. Those sitting in the regulatory bodies don't do their work properly, due to a lack of efficiency, lack of incentives, and corruption. (quoted in Sattar, 2013)

Similar insights come to the fore when analysing other building collapses in the same region. For instance, on 27 September 2013 a five-storey building collapsed in the Mazgaon Area of Mumbai, India. Over 60 people died. Again, illegal building activities appear as the cause of the collapse, while enforcement was found to have failed. Officials did not act on reports that highlighted the structural lack of safety of the building (Cook et al., 2013; Narayan and Jain, 2013; Purohit, 2013).

The earthquake that hit the Padang region of Sumatra, Indonesia, on 30 September 2009 destroyed about 100 000 houses and 1000 other buildings, and killed more than 700 people. After the earthquake the head of Sumatra's Building and Environmental Management Department stated:

> There are problems with construction quality. There are rules that buildings must be built by certified entities, but enforcement has been less than strict. (Quoted in the *Brunei Times*, 2009; on lacking enforcement practice of construction codes in this case, see also EERI, 2009; Gunawan, 2009)

Such a lack of enforcement practice, and enforcement capacity in developing economies more generally, has been highlighted before (Hettige et al., 1996; Kirkpatrick and Parkers, 2004; Nath and Behera, 2011). It would, however, be a misinterpretation to consider it a problem that just affects developing economies.

On 24 April 2003 a balcony snapped off the top floor of a recently occupied condominium building in the city of Maastricht, the capital of a

Dutch province. The owners of the unit happened to be outside on their balcony when the incident occurred. They were killed instantly by 50 tons of falling rubble (De Volkskrant, 2003). The balcony incident quickly became a media hype in the Netherlands, and not before long the media uncovered reports of failing construction code enforcement practice by the local construction authorities (NRC, 2003). The media found that when the building was under construction the technical design of the balconies was changed at least five times, 'not taking into account the structural aspects too much' (Cobouw, 2003). Research by the ministry responsible for the development of the Dutch construction codes noted that the municipal building authority lacked information to assess building plans against Dutch building regulations, that the building was not built according to the building permit issued by the municipality and that the local construction authority noticed these deviancies but did not act (VROM, 2003). After inspecting the building under construction they noted in their inspection reports: 'no comments' (VROM, 2003, chapter 8).

These cases echo experiences of lacking construction code enforcement practice from all over the world. In Australia (Chivell, 2005), Chile (Margaret, 2010), France (Starossek, 2006), Germany (Der Spiegel, 2008), Haiti (Bhatty, 2010), Italy (Povoledo, 2009), Japan (The Japan Times, 2005), Nigeria (Reuters, 2013), Turkey (Lynch, 2011), the United Kingdom (Imrie, 2004) and the United States (Davis, 2008) similar shortfalls have been reported. This ever-growing list of examples of failing construction code enforcement indicates that enforcement is not a trivial matter. The effectiveness and efficiency of direct regulatory interventions stand or fall by it.

2.3.2 Reforms of Construction Code Enforcement

The problem of construction code enforcement is twofold. On the one hand, it relates to the organization of enforcement, on the other, to the operationalization of enforcement (Van der Heijden, 2009).

In terms of organization, the enforcement of construction codes and related direct regulatory interventions, is often organized on a local level. Municipal building authorities assess applications for building permits against construction codes, inspect construction work and inspect buildings when in use – at least, in an ideal world.

In the real world these authorities, however, often find themselves limited by strict budgets. They can only hire a small number of staff that is restricted in what it can do in terms of building code enforcement. In addition, when hiring staff an important choice has to be made: What specialism to hire? Someone with expertise in the energy performance of buildings? Someone who knows about structural safety? A generalist maybe?

Being restricted in the number of staff that an authority can hire also implies that this authority is restricted in what enforcement actions it can carry out. With an oversupply of enforcement work, and an understaffed workforce, building authorities are likely to prioritize particular enforcement actions over others (Van der Heijden, 2009). For instance, high-risk commercial buildings will be inspected in more depth than low-risk residential buildings; or the work of a builder with a reputation of being somewhat dodgy will be inspected in more depth than the work of a builder who holds an ISO 9001 quality management system certification. Such a risk-based approach to enforcement inevitably leaves some risks uncovered (Baldwin et al., 2000; Flüeler and Seiler, 2003; Hood et al., 2001; Spence, 2004).

In terms of the operationalization of enforcement, construction codes and the urban environment more generally provide particular complications to building authorities. Buildings and other infrastructure projects are normally unique in their design, the bringing together of different technologies and the working together of specific project teams. It takes a long time to construct a building, and this is mostly done on-site under uncontrollable conditions. This makes it impossible to fully assess a construction work. This is especially true when it is understood that building plans only show how the work is supposed to be constructed; during construction an inspector cannot be present to inspect each and every action taken; and once finished much is hidden behind walls, ceilings and floors. Compare this, for instance, with the construction of a car, which is manufactured under controlled conditions in a factory, within a relatively short period of time, and each and every aspect of the car can be dismantled for inspection if need be (Mikler, 2009).

Seeking to overcome the problem of the organization of enforcement, many countries around the world have privatized construction code enforcement. Although various forms of private construction code enforcement exist (Pedro et al., 2010), normally a private organization or individual can be registered as a private inspector when meeting criteria related to education, working experience and professional indemnity insurance (Van der Heijden, 2009). It is often expected that private inspectors will specialize in specific building types or construction activities, such as hospital specialists or fire safety specialists. The lack of staff often makes this type of specialization impossible for smaller municipal building authorities.

Recent research indicates that privatized construction code enforcement comes with advantages and disadvantages (Imrie, 2007; Van der Heijden, 2009). In countries such as Australia and Canada private inspectors have indeed become specialized in specific building types or construction activities, which has overall resulted in better trained and better skilled inspectors than under a pre-privatization situation. Better trained and better skilled inspectors are considered to be more effective in their enforcement work, which may result in higher levels of compliance with construction codes. However, privatization is considered to come with its own risk as well. When those governed (that is, building owners, contractors) pay their own inspector for enforcement activities a danger of corruption exists. In my own studies on the topic private inspectors were repetitively referring to anecdotal evidence of other inspectors traversing on the edge of being strict in enforcing construction codes and cutting corners. Without an additional layer of oversight on these private inspectors a privatized system of construction code enforcement is likely to fail (see also S. Yu et al., 2013).

In both Australia and Canada private inspectors were added to the existing system of local building authorities. As such, those governed have a choice of involving a private inspector or a public authority in their construction projects. My research indicates that different groups of people have different expectations from their interactions with construction code enforcers – both governmental authorities and private inspectors.

Professionals in the industry, such as architects, engineers and developers, expect a specialist service for the complicated works they carry out. These professionals have, normally, repeated interactions with construction code enforcement since designing, constructing or managing buildings is their day-to-day job.

Non-professionals in the industry, predominantly homeowners, expect to be guided through the construction code enforcement process for the, generally, relatively simple works they are involved in, for example, adding a carport to a house or installing solar panels.

Distinguishing those governed by direct regulatory interventions in professionals (or repeat players) and non-professionals (or one-shotters) may help to develop governance approaches targeted to each of these group's needs (terminology from Hirschman, 1970). Such targeted governance approaches may very well achieve better outcomes in terms of compliance than the current one-size-fits-all building regulatory enforcement regimes that are implemented by governments all over the world (see Van der Heijden, 2010c, 2011, 2013a).

2.4 CONSIDERATIONS

The real-world examples discussed in this chapter confirm the account developed from reviewing the extant literature on direct regulatory interventions. For governing urban sustainability and resilience, one of the major strengths of direct regulatory interventions is that they are backed by law. This implies that governments can use force, if necessary, to ensure compliance with these interventions. This holds, most clearly, for statutory regulation.

The other approaches to governance discussed in this book (see Chapters 3 and 4) lack this specific characteristic. Besides, governmental regulation may be considered as more legitimate, by the general public, than voluntary programmes and market-based governance tools developed by businesses, civil society groups and individuals. The general public may question whether such 'self-regulated' actors are truly acting in the public interest or whether they act to seek their private interests served (Bovens, 1998; Verhoest et al., 2007).

2.4.1 Some Concerns About Direct Regulatory Interventions

Direct regulations may, however, be critiqued for seeking bottom line compliance only. That is, those subject to these regulations will likely consider compliance with these regulations as sufficient, and are unlikely to achieve 'beyond compliance' levels (Van der Heijden and De Jong, 2009). Also, although it is often assumed that increased regulatory requirements will stimulate significant innovation in the building sector, empirical research points in the other direction. It is more likely that conventional building materials and technologies will be improved to meet new requirements (Beerepoot and Beerepoot, 2007).

Subsidies, taxes and other economic incentives, such as tradable permits, may overcome these issues. The strength of these tools is that they seek to change behaviour by not only focusing on what is not allowed but by using financial incentives to steer those governed towards behaviour that is desired.

Those governed are incentivized to reduce potential harms to zero instead of to the specified standards (Baldwin et al., 2011). For instance, a subsidy for solar panels may incentivize a household to install enough solar panels to fully generate all the energy it needs.

Subsidies may, however, be critiqued as well for having too limited effects. The same household may decide to not participate in the subsidized solar panel plan. Subsidies can even be critiqued for resulting in negative effects when, for instance, undesirable behaviour is (indirectly) funded. The example of the National Flood Insurance Plan in the United States is telling.

A final critique to subsidies is that they may only be accessible to the already well off, or those who understand how to apply for subsidies. Again, the example of a household that wishes to participate in a subsidized solar panel plan is telling. The household likely needs to have the funds to pay the upfront costs of the solar panels, or at least needs to have the funds to pay the non-subsidized costs for their solar panels. Subsidies and tax deductions alike may distribute taxpayers' money unequally among society, as some of the examples in this chapter show.

The major weakness of direct regulatory interventions is that they require significant enforcement efforts to ensure compliance (for a more general discussion on compliance and enforcement, see Parker and Lehman Nielsen, 2011). This may make this particular approach to governance costly and time consuming, but it would be unfair to only judge direct regulatory interventions as such; the other approaches to governance (discussed in Chapters 3 and 4) come with their own enforcement dilemmas. Particularly the enforcement of statutory regulation is complicated in urban settings because much of the regulated entities are 'hidden' behind walls and ceilings, under pavements and in concrete. Further, because buildings and infrastructures are often constructed on-site it is difficult to guarantee (and inspect) a constant quality of work. In terms of the increasingly popular performance-based and goal-oriented standards, it may be argued that these result in further enforcement complications when it is unclear to those governed as to what complies and what not, or when it is unclear to those enforcing regulations as to what can be accepted as compliant and what not (Van der Heijden, 2010a).

Local governments are, mostly, the responsible authority for the enforcement of statutory building regulation. They often face a lack of enforcement capacity, both qualitative and quantitative, to ensure high levels of enforcement of building regulation (Van der Heijden and De Jong, 2009). Responding to this issue, governments around the world have often privatized this enforcement. Privatized building regulation may indeed partly solve this lack of capacity. However, privatized enforcement is no guarantee for success. An additional layer of oversight on private construction inspectors may be needed to ensure that they operate in the public's best interest.

Besides, especially in technologically advanced areas, such as urban sustainability and resilience, governments may lack the knowledge to set meaningful standards (Gunningham and Grabosky, 1998); and the relatively slow development process of setting standards may result in situations where novel technologies are not yet accepted as complying with standards (Baldwin et al., 2011). It is therefore hopeful to see that some governments are now giving particular attention to innovative solutions that seek higher levels of compliance, or even fast-track proposals for highly sustainable buildings (for example, Green Door in Queensland, Australia). It is further hopeful to see that governments are actively working together with business and civil society representatives in developing statutory building regulation (for example, the role that DIN plays in the development of building regulation in Germany).

Nevertheless, this approach to governance may also result in legalism where the process of developing and implementing direct regulatory interventions apparently becomes more important than their outcomes. Regulators are sometimes found to over-regulate and introduce too complicated and over-inclusive rules, and haphazardly respond to public demands and introduce too general and broad-brush regulations (Bardach and Kagan, 1982). With extreme weather events having more negative impacts on cities and other urban areas, it is to be expected that policy makers want to quickly react to floods, hurricanes and bushfires.

In analysing direct regulatory interventions it appears that policy makers like to single out a problem and introduce a solution to that problem. For instance, the energy performance certificates in Europe were introduced to overcome information asymmetries. Problems that stand in the way of improving the environmental sustainability and resilience of buildings, however, are often too complex and too interrelated to identify and solve with a single governance tool. It is more likely that a combination of tools is successful in addressing problems (Beerepoot and Sunikka, 2005; Driesen, 2006; Gunningham and Grabosky, 1998; Lee and Yik, 2004; Stewart, 2006). Finally, once a direct regulatory intervention is in place it appears difficult to remove. Despite years of deregulation efforts that exactly sought to free the market from direct regulatory interventions, there now seem more governmental rules and regulatory tools than ever before (Majone, 1993; Vogel, 1996).

2.4.2 Potential to Overcome the Three Main Governance Problems

Whether and how are direct regulatory interventions able to overcome the three main governance problems discussed in Chapter 1? While the problem caused by the grandfathering of existing buildings and infrastructure is chiefly related to this approach to governance, it is not to say that governments are unable to solve it through direct regulatory interventions.

Subsidies, tax breaks and regulatory relief such as the fast-tracking of particular building design proposals may incentivize building developers and owners to seek higher levels of compliance of their buildings than that required by governmental regulation. Linking building regulations with insurance requirements may incentivize developers and buildings' owners to improve the resilience of their buildings, especially when the fees for non-resilient buildings face a significant increase.

The problem of the time lapse between urban development and regulatory response in rapidly developing economies appears so intertwined with this approach to governance that any of the available tools are unlikely to solve it. Here governments from developed economies may be in the best position to share their experiences with governments in developed economies, and to support them in developing adequate regulatory responses (Chapter 3 discusses a series of examples of how this is currently being done).

In addition, in terms of implementation and enforcement, sustainable building regulation may provide a new dilemma (in developed and rapidly developing economies). Where building authorities may historically be considered to have the legitimacy to implement building regulation that addresses public health and safety and issues, and use force to enforce these, it is questionable whether the public in general will accept the implementation of far-reaching building regulation that seeks to address environmental sustainability (Hirokawa, 2009). This may hold all the more true in rapidly developing economies where far-reaching requirements in terms of achieving urban sustainability may negatively affect or slow down (short-term) economic development.

Finally, direct regulatory interventions may in part be an adequate response to the wicked set of market barriers Chapter 1. Direct regulatory responses have, after all, in the past proved successful in overcoming market barriers and in achieving desired collective ends such as cities that are safe from rapidly spreading fires. But, because of the above-mentioned considerations, that is, grandfathering, the time lapse faced in rapidly developing economies and issues related to enforcement of such interventions, not too much should be expected from this approach to governance in overcoming the wicked set of market barriers. It is more likely, as Chapters 3 and 4 explain, that governance tools that are developed by, or in collaboration with, business and civil society representatives are more effective in overcoming this wicked set of problems. Partly because these approaches to governance are more aware of utilizing the knowledge of business and civil society; and partly they may be considered less intrusive, less paternalistic and more legitimate by businesses and civil society groups and individuals.

2.4.3 Promises for Governing Urban Sustainability and Resilience

In conclusion, given governments' decades (to centuries) of experience with direct regulatory interventions it may be expected that there is sufficient institutional capital, especially in developed economies, to address urban sustainability and resilience through this mode of governance to a certain extent.

Introducing statutory regulation for urban sustainability is a powerful tool (Shapiro, 2009) and often suggested as one of the most important tools in achieving an environmentally sustainable built environment (UNEP, 2006). Governments may be able to implement direct regulatory interventions on a large scale, and have the power to achieve compliance, if necessary by force. Also, in terms of governing urban resilience, experiences with the regulation (and enforcement) of health and safety of the built environment may be valuable for developing and implementing governance tools that seek to further improve urban sustainability.

That having been said, the high implementation costs, and in particular the costs related to enforcement, of this approach to governance should not be underestimated (Lee and Yik, 2004). Also, those same decades of experience have indicated other shortfalls of direct regulatory interventions such as the difficulty in setting timely meaningful standards that optimally use available and future technologies, are enforceable and do not over-burden citizens and businesses.

3. Collaborative governance

Since the 1990s the shift from 'government to governance' (Rhodes, 1997) has been presented as a solution for the 'unreasonableness' of direct regulatory interventions (Bardach and Kagan, 1982). This shift reflects both ideological standpoints and empirical evidence that governments are (and should) no longer be the sole decision-making authority in, for instance, urban sustainability and resilience.

The call for a sharing of decision-making powers and collaboration in governing fits a longer trend of a changing relationship between government, businesses and civil society that includes other trends such as deregulation (Majone, 1990; Vogel, 1996), privatization (Abramovitz, 1986; Hodge, 2000) and new public management (Hood et al., 1998; McLaughlin et al., 2002).

Collaborative governance[1] appears particularly popular in addressing contemporary urban problems (for example, Beatley, 2000; Betsil and Bulkeley, 2006; Evans et al., 2005; Gruvberger et al., 2003; Hoffmann, 2011; McManus, 2005; Newman et al., 2009; Nijkamp and Opschoor, 1995; Portney, 2003). This chapter seeks to better understand this trend of collaborative governance and its promises for achieving urban sustainability and resilience. It first briefly introduces the ideological and emerging empirical literature on collaborative governance, and discusses its promises and potential risks. It then seeks to understand whether these expectations are met in a series of real-world settings where governments, businesses and civil society groups and individuals collaboratively seek to achieve urban sustainability and resilience by developing and implementing innovative governance tools.

3.1 CHARACTERISTICS OF COLLABORATIVE GOVERNANCE

Direct regulatory interventions, such as those discussed in Chapter 2, have a fairly straightforward development and implementation process and structure. This is not the case for collaborative governance. The literature discusses a wide variety of processes in which various actors are involved. The tools related to collaborative governance are also less strictly defined than those related to direct regulatory interventions. The literature is further complicated since the dividing line between ideological debates and evidence on collaborative governance is sometimes thin. The discussion that follows therefore has a focus on the tension between normative discussions and insights derived from empirical studies.

3.1.1 Actors Involved

Key to collaborative governance is the working together of governments, businesses and civil society groups and individuals in governing. This is a considerably different approach to governing than through direct regulatory interventions in which non-governmental actors are predominantly considered to be mere subjects to governmental interventions.

The role of non-governmental actors in governance

Why would governments collaborate with non-governmental actors in the development and implementation of governance tools, and vice versa?

It is generally assumed that non-governmental actors have better knowledge of their day-to-day behaviour and that of their peers than governments can ever obtain. For example, a builder is best aware of its own practice, knows how to deal with construction codes and will likely have a clear idea about which regulations are easy to comply with and which ones are not. Also, in the example of private sector building inspectors discussed in the previous chapter (Subsection 2.2.3), private sector inspectors were very much aware of the work of their peers. They could tell which of their peers carried out strict regulatory enforcement and which of their peers were more lenient in doing so (Van der Heijden, 2009). Including this knowledge of non-governmental actors in the development of governance tools is expected to result in more suitable and effective governance tools than could be delivered by (somewhat) distant bureaucrats (Croci, 2005; Hendriks, 2009; Karkkainen, 2004; Lobel, 2004; Scott et al., 2004; Solomon, 2008).

Collaboration is also expected to result in more efficient governance tools since resources can be allocated to better suit those governed (Baldwin et al., 2011; Fairman and Yapp, 2005). Including non-governmental actors in the development and implementation process of governance tools is expected to result in a higher willingness of these actors to comply with these tools once in force. They may feel ownership and responsibility for achieving the goals of these (Ansell and Gash, 2008; Blackstock and Richards, 2007; Lobel, 2004), which they may lack when subject to more traditional approaches to governance such as those discussed in Chapter 2 (Bodansky et al., 2008; Holley et al., 2012). At the same time, collaboration may help non-governmental actors in that they can try to have governance tools implemented that are least intrusive to them (Maxwell et al., 2000; Reid and Toffel, 2009).

In addition, collaboration is expected to result in increased legitimacy and accountability of governance tools. Those governed may feel empowered by the collaborative governance process. It is expected that through collaboration those governed will consider these tools as more legitimate when they were heard in their development process. Also, because the tools are developed in collaboration, participants may feel a shared responsibility for the outcomes of these tools, which in turn may improve their accountability (Lobel, 2004; Noveck, 2011; Sabel and Simon, 2006; Wilkinson, 2010).

The specific roles of non-governmental actors in collaborative governance vary. They can be far-reaching and overlap with the roles of governmental actors as discussed in Chapter 2, such as the development, implementation and enforcement of particular rules. It goes without saying that non-governmental actors do not have similarly far-reaching enforcement powers as governments do – that is, they cannot rely on strict disciplinary actions such as issuing fines or imprisoning violators.

The role of government in collaborative governance

Governments are expected to take up particular roles in collaborative governance. Some of these relate to the more traditional roles discussed in Chapter 2, while other roles are less traditional. As will become clear throughout this chapter (and also in Chapter 4), in achieving urban sustainability and resilience through collaborative governance city governments are especially active in taking up such non-traditional roles (see also Evans et al., 2005; Newman et al., 2009).

These city governments may act as initiator or leader of collaborative governance processes. Such an initiating or leading role is considered to help (potential) participants to find one another, to merge diverse interests and to ensure that a collaborative group will reach relevant and effective solutions (Davis, 2002; Sabel et al., 2000). Such leading or initiating manifests itself, for instance, in the development of platforms, centres and networks (Steurer, 2010). City governments may further act as assembler of different collaborative governance processes. By doing so, they can seek to ensure their cohesion among different actors and different governance tools (Davis, 2002).

Another essential role of city governments may be to act as guardian of collaborative governance tools. It is often found that without meaningful enforcement these are unlikely to achieve desired outcomes (Bailey,

2008; Lyon, 2009; Rivera and de Leon, 2004). Governments are well suited to provide enforcement capacity, meant to ensure that participants fulfil their obligations (Gunningham, 2009).

Finally, city governments may act as supporters or facilitators of collaborative governance processes, which are initiated by non-governmental actors. Support may be provided through positive incentives, such as financial or organizational support, or through negative incentives, such as the threat of legislation coming into place if not enough participants join collaborative governance processes or if they do not comply with those in which they participate (Héritier and Eckert, 2008; Hertier and Lehmkuhl, 2008).

3.1.2 Governance Tools

The tools of collaborative governance are less contained than the direct regulatory interventions discussed in Chapter 2. Various typologies are presented in the literature, but all have a different focus and different boundaries (for example, Ansell and Gash, 2008; Glasbergen et al., 2007; Steurer, 2013; Wurzel et al., 2013). In this chapter the focus is on covenants and (negotiated) agreements, and partnerships and networks (Delmas and Young, 2009; Steurer, 2013).

Negotiated agreements and covenants

The most well-known examples of collaborative governance come in the form of negotiated agreements and covenants (Andrews, 1998; Bressers et al., 2009; Potoski and Prakash, 2005a).

Under a negotiated agreement or covenant an individual, a firm or a group of individuals or firms pledge to achieve a particular goal and government in return commits itself to a related objective, for instance, supporting the private sector actors in achieving their goal or not introducing any regulation during a specific time span of the agreement or covenant (Seyad et al., 1998; Ten Brink, 2002).

The current literature is not univocally positive about the effectiveness and efficiency of negotiated agreements and covenants in achieving societal goals (for a recent review of the literature, see Chittock and Hughey, 2011). In particular, it finds that without credible targets, ongoing monitoring of performance and a credible threat of enforcement these agreements and covenants are unlikely to achieve desired results. This reflects many of the insights about the strengths and weaknesses of the direct regulatory interventions discussed in Chapter 2.

Partnerships and networks
Partnerships and networks are another way for private sector actors and governments to collaborate. Throughout this book partnerships and networks are understood as less coercive forms of governance than negotiated agreements and covenants. They bring together actors to share information, learn from each other and pool resources to test or pilot specific governance tools.

The strengths of such partnerships and networks are found in the ability of governments, businesses and civil society groups and individuals to develop solutions for societal problem in collaborative and consensus-oriented processes (Ansell and Gash, 2008; Lewis, 2011; O'Flynn and Wanna, 2008). In addition, such partnerships and networks can, often, be developed in a relatively short timeframe, and because their level of coercion is low they may be attractive to a wide range of prospective participants.

3.1.3 Development and Implementation Process

Collaborative governance as a process to develop and implement governance tools is both lauded and critiqued in the current literature. It is often found that its ideological foundations are difficult, if not impossible, to realize in real-world settings.

One ideal of collaborative governance is particularly critiqued: giving all relevant stakeholders voice in collaborative governance processes (NeJaime, 2009). Yet, how to realize this ideal of involving all relevant stakeholders? After all, the larger the number of actors involved in the development and implementation of governance tools, the more difficult it will be to reach a consensus. It could be argued that the extent of collaboration and deliberation as 'suggested' in the literature will inevitably face practical constraints related to collective action problems (Olson, 1965).

A second problem relates to the ambiguity of the roles of the actors involved in these governance processes (Dietz et al., 2003; Savan et al., 2004; Singleton, 2002). Achieving a meaningful sharing of power between stakeholders appears to be extremely difficult in real-world settings (Arnstein, 1969; Noveck, 2011). All too often governmental actors are found to allow non-state actors to participate in decision making, but to not include their voice in the implementation of governance tools (Dryzek, 2005; Sunstein and Hastie, 2008). Collaboration may in practice result in a situation where 'talk is disconnected from power' (Noveck, 2011, p. 89). This, for instance, takes into consideration whether information is openly available and readily shared or whether a rational argumentation and dialogue between citizens, professionals and public officials is enabled (Barber and Bartlet, 2007; Gutmann and Thompson, 2004; Webler and Tuler, 2006; Wiklund, 2005).

The extant literature is also sceptical about the possibilities stakeholders have for meaningful and broad participation in collaborative governance processes (Ford and Condon, 2011) and the extent to which these may achieve distributive justice (Alexander, 2010). Due to an inability to participate in collaborative decision making, the voices of weaker stakeholders are rarely included in the outcome of such processes (Eversole, 2010; Scott, 2009). In such situations collaborative governance may become prone to the manipulation of stronger stakeholders (Ansell and Gash, 2008; Noveck, 2011).

In a related vein, the actual impact of collaboration on accountability is unclear. The shared responsibility for the outcomes of collaborative governance processes may turn out to be problematic in a situation of undesired outcomes (Pollak and Slominski, 2009; Smismans, 2008; Wilkinson, 2010). No one is then likely to take responsibility for these outcomes; a 'problem of many hands' (Thompson, 1980).

3.2 FROM THEORY TO PRACTICE: URBAN EXPERIENCES WITH COLLABORATIVE GOVERNANCE

The variety of partnerships, networks, covenants and (negotiated) agreements between governments, the private sector and civil society appears immeasurable (Backstrand et al., 2010; Hickson, 2009; Hoffmann, 2011). To give an illustration of collaborative governance for urban sustainability and resilience, this section discusses a number of examples and experiences related to the different approaches to collaboration identified. The focus is, as in Chapter 2, on examples that seek to improve the sustainability and resilience of buildings. It first addresses a series of government to government collaborations (Subsection 3.2.1), and then discusses a series of collaborations that are led by governments (Subsection 3.2.2), by businesses (Subsection 3.2.3) and by civil society groups (Subsection 3.2.4).

3.2.1 Government to Government Collaboration

City networks: creating and communicating knowledge

Government to government collaboration can be uncovered on different levels. The Ontario Regional Adaptation Collaborative in Canada operates on a regional scale. Under this collaborative the governments of Canada and Ontario seek to support communities to adapt to climate risks. The aims of this collaborative are knowledge generation, the development of governance tools and building partnerships (Government of Ontario, 2011).

The collaborative has resulted in a number of projects that address the management of extreme weather risks, water management and community development planning. The collaborative takes a hands-on approach. It organizes training programmes for staff from Ontario's local governments to better understand the potential impacts of climate change on local communities and how to respond to these. It further aims to document and communicate experiences of how communities seek to adapt to climate change and the lessons they have learned in doing so (Clean Air Partnership, 2012).

The European Future Cities Network operates on an international level. This collaborative of nine cities in the United Kingdom, the Netherlands, Belgium and Germany seeks to make city regions in Northwest Europe fit to cope with predicted climate risk impacts. The participants cooperate to develop, apply and improve assessment criteria for urban sustainability and resilience, with a focus on existing infrastructure. The collaborative receives financial support from the European Union. Findings are shared among the participants and communicated to a wider audience through a series of conferences and series of project reports (Future Cities, 2013).

On an even larger scale the Mexico City Pact is a network of 290 cities in 60 countries, spanning five continents. These cities have committed to mitigate greenhouse gas emissions, as well as to implementing community adaptation measures. Through the pact the participating cities develop best practices and share lessons learned (Global Cities Covenant on Climate, 2013).

Perhaps the most well-known and best documented examples of government to government collaboration are the international networks ICLEI – Local Governments for Sustainability, a network of more than 1000 cities and local governments that work together to achieve urban sustainability and resilience, and the C40 Cities Climate Leadership Group, a network of the world's largest cities that collaborate to reduce greenhouse gas emissions through increased urban sustainability.

Through such networks cities share information and knowledge, experiment with local governance tools to improve the sustainability or resilience of their building stock or compete for being the most sustainable or resilient city. Both networks have achieved considerable successes, and a number of success criteria are mentioned in the literature

(Betsil and Bulkeley, 2007; Bhagavatula et al., 2013; Hoffmann, 2011; Reams et al., 2012; Rosenzweig et al., 2010).

First, both networks are built around outreach, education efforts and learning. Through participation, cities have a wealth of knowledge to access. The scale of the networks means that experiments with governance tools can be carried out simultaneously in different cities, but overseen or even conducted by a single set of researchers. Due to their scale these networks have sufficient funds to attract professional researchers or have data that is attractive enough for academics to voluntarily participate in their research projects. The professionalism of data collection and data analysis has resulted in high-quality research results that are communicated in a highly accessible manner.

Second, both networks have set clear goals for participation, but keep participation criteria low. Cities commit to improving their environmental performance and are supported in doing so by the networks. Yet, participating cities are not disciplined for not achieving the goal to which they have committed. This combination of relatively low participation criteria and relatively low levels of enforcement is often found to result in high levels of participants (Potoski and Prakash, 2009). However, these relatively low participation criteria are criticized. Through participation in the networks cities may give the illusion of addressing issues related to sustainability and resilience, while *de facto* they do not commit to any strict target (see discussions in Hoffmann, 2011). In addition, government to government collaboration may run the risk of becoming networks of pioneers for pioneers when lessons learned are only communicated with participants (Kern and Alber, 2010).

Third, the networks have significant visibility. They have attracted the world's major cities, which may give the networks as a whole a high level of trustworthiness. Because of the participation of the world's major cities these networks also have significant political voice, which becomes evident, for instance, in their participation in the United Nations international climate negotiations. In particular, the role of mayors of these cities appears relevant. They are highly experienced with complex policy processes and know-how to influence and set policy agendas. It is telling that in 2014 UN Secretary-General Ban Ki-moon appointed former New York Mayor Michael Bloomberg as his special representative for cities and climate change (Nichols, 2014).

A pearl in the Jawaharlal Nehru National Urban Renewal Mission

The Jawaharlal Nehru National Urban Renewal Mission (JNNURM), India, provides for another interesting government to government collaboration. This city modernization scheme was launched by the

Government of India under the Ministry of Urban Development in 2005. The JNNURM seeks to improve urban governance, the quality of life and the quality of infrastructure in Indian cities and is regarded as one of the most ambitious contemporary policy programmes in India. The Indian Government looked to invest roughly US $20 billion into the urban sector development of 65 selected cities, over an initially seven-year period. The JNNURM was later extended by another two years (for more on the JNNURM, see Jain, 2010). The JNNURM is unique and ambitious for India in that it allows local governments significant freedom in achieving the scheme's goals.

The JNNURM is both lauded for its ambition and critiqued for its actual performance. An audit of its performance in 2012, by the Comptroller and Auditor General of India (CAP, 2012), highlighted that many original goals have not been accomplished (Grant Thornton, 2011). For instance, out of the intended 1.6 million planned dwelling units, only 26 per cent was realized by the end of the JNNURM's first seven-year cycle and only half of those dwellings were occupied. The Comptroller and Auditor General critiques the Ministry of Urban Development for not having in place an adequate administrative apparatus for handling a scheme of the magnitude of the JNNURM. It particularly advises the Government of India to put in place stronger enforcement of the scheme to prevent 'diversions [of funds] to in-eligible beneficiaries' and to 'introduce a zero tolerance policy at all levels in respect of irregular expenditure and division of funds by way of greater financial discipline' (CAP, 2012, pp. vii–viii).

What makes the JNNURM of particular interest is that it has resulted in a Peer Experience and Reflective Learning (PEARL) Network, India's first city governance network (Grant Thornton, 2011; Vaidya et al., 2010). The PEARL Network is a typical example of a government to government collaboration in which the 65 selected cities seek to exchange knowledge.

PEARL was launched in 2007, supported by the Ministry of Urban Development and the World Bank. One of the most innovative aspects of PEARL is the twinning of cities. This is a concept in which a better performing city is paired with a less well-performing city, aiming to transfer the lessons learned from the former city to the latter. PEARL further collects and communicates findings through a regular newsletter, documents best practices in reports, regularly organizes workshops and seminars for participating cities and maintains a website that further communicates lessons about the implementation of the JNNURM (PEARL, 2010).

The best practice reports discuss the process, outcomes and transferability of a particular practice, which has proved effective in a particular context. They do so by discussing, in a codified manner, the goals of the practice, its budget, the relevant stakeholders involved, the problems encountered and the impact of the best practices. By the end of 2013 a total of 133 best practices were available from the PEARL website (PEARL, 2013). A financial incentive scheme was in place in the first years of PEARL to ensure that participating cities would actually provide information on their experiences with the implementation of the JNNURM. As a result, PEARL now has a database of experiences, lessons about successes and failures, and best practices that by the end of 2013 had been accessed over 45 000 times (PEARL, 2013).

One of the key successes of PEARL is the broad and easy availability of information relevant to city officials, as well as the leading role one of the initiating administrators took (Maher, 2012; Vaidya et al., 2010). This individual actively addressed municipalities to feed the PEARL website with best practices and lessons learned, and ensured that funds were available in the first year to reward municipalities for providing such information. Due to its success it is likely that the network will be sustained beyond the implementation of the JNNURM (Maher, 2012).

3.2.2 Government-led Collaborations

City planning: citizen participation and political swings

All over the world city governments involve citizens in the development of city plans (Brabham, 2009; Derr et al., 2013; Evans-Cowely and Hollander, 2010). An example is Growing a Green Heart Together, the Australian city of Brisbane's community plan.

The plan expresses Brisbane's ambitions in becoming Australia's most sustainable city, and to have its city council carbon neutral by 2026. The city considers that this can only be achieved if individuals, households, schools, businesses, industry and the city council work together (City of Brisbane, 2009). With this vision in mind, the plan was developed in consultation with the community through meetings and community forums with a range of stakeholders. Various activities were organized to get a broad group of citizens involved in the development of the community plan. Citizens were invited to draw their vision or describe it in words. A web portal was launched to inform citizens and to allow them to share their ideas on the future of Brisbane. The administrators involved in this collaborative process especially valued these different approaches to get citizens on board. They considered the low participation threshold provided, for instance, anonymously through the web portal, combined with more formal and traditional participatory approaches such as discussion sessions a successful combination (Van der Heijden, 2013b, 2013d).

In order to make the implementation of the plan a success a number of additional governance tools was implemented. Two competitive grants were launched in 2009 to stimulate developers and owner-builders to construct innovative buildings with high levels of environmental performance. Unfortunately, the two grants were terminated in 2011. As a result of the 2011 flood damage (see Chapter 2), the Brisbane City Council decided that funding for these grants needed to be redirected towards Brisbane's flood recovery effort. In addition to these grants, homeowners were addressed through the ClimateSmart Home Service. This was a service funded by the council to have residential buildings inspected and to provide their owners or users with information on how to achieve reductions in energy and water consumption. The service was terminated in April 2012 by the Liberal-Conservative Party. They took over political power from the Liberal Party in the state of Queensland, of which Brisbane is the capital, in March 2012 (Hurst, 2012). It considered the service a too-large tax burden on Brisbane's citizens.

This political change in Brisbane appears to have resulted in a significant swing in how the past and current city and state governments address sustainability. In a period of less than seven years the idea of urban sustainability has been addressed in various ways.

In 2006 the then city plan addressed the concept as: 'To apply sustainability, we need to bring together economic, environmental and community considerations' (City of Brisbane, 2006, p. 11).

In 2009 the Growing a Green Heart Together community plan does not refer to economic growth or economic prosperity. It is fully built around the idea that Brisbane is to become the most sustainable Australian city and to have its council carbon neutral by 2026 (City of Brisbane, 2009).

In 2013 a new city plan was issued, building on the earlier 2006 city plan. Strikingly, the 2013 city plan is introduced by the city's then Liberal-Conservative government stating that 'Brisbane City Council is committed to facilitating economic growth and maintaining prosperity in Brisbane through sustainable development' (City of Brisbane, 2013a), while the plan is not referring to the earlier ambitious goal of the city in terms of carbon emissions or aiming to having its council carbon neutral by 2026. The new plan appears fully focused on economic growth, and pays very limited attention to environmental and resource sustainability.

Intriguingly, for both the 2009 community plan and the 2013 city plan citizen participation were important parts of their development (for public participation in the 2013 city plan, see City of Brisbane, 2013b). It is somewhat striking to see such different directions in the two plans within such a short period of time. Have the citizens of Brisbane en masse significantly changed their opinion about environmental sustainability and carbon neutrality of their city council? In the 1960s Sherry Arnstein (1969) had already noticed that in citizen participation talk is often disconnected from action. In analysing comparable examples of citizen participation she noticed that citizens are often heard in participatory processes, but that policy makers do not reflect their input in policy documents. With the significant differences in the 2009 Brisbane community plan and the 2013 city plan, one wonders whether Arnstein's insights also hold for either or both of the participatory processes that led to these plans.

Negotiated agreements and covenants

A good example of covenants and negotiated agreements in urban environments are the Climate Change Sector Agreements between the state government of South Australia and business entities, industry sectors, community groups and regions (Government of SA, 2007).

These agreements help the state government to achieve their ambitious goal of reducing greenhouse gas emissions to an amount that is less than 40 per cent of 1990 levels by 2050. Among others, agreements have been reached with the South Australian commercial property sector association, an investment bank and the local cement sector.

The agreement with the commercial property sector association, for instance, aims to promote a national recognized building benchmarking tool for existing commercial buildings in the industry (the voluntary National Australian Built Environment Rating System, NABERS; discussed further in Chapter 4). By benchmarking buildings insights can be given into their environmental performance. The agreement further sought to develop and implement educational and promotional strategies to encourage property owners and tenants to improve their buildings' environmental performance. The agreement expresses the importance of developing best practices and communicating these throughout the sector.

The commercial property sector association was responsible for collecting and communicating data, for consulting with property owners and tenants, and for supporting these in improving their environmental performance. In return, the Government of South Australia financially and administratively supported the actions undertaken by the commercial property sector, pledged to publicly acknowledge the achievements of the sector association and its participants, and committed to realign various government policies and programmes that were of interest to the property sector (Government of SA, 2009b).

Reviews of these agreements are largely positive (Government of SA, 2009a; LGA, 2013). Administrators of these agreements referred to the publicity that was given to the agreements by the local media and the public availability of the arrangements on the internet as two of their main success factors. They thought all parties involved in the agreements felt pressure to comply with these because of the public commitments made (Van der Heijden, 2013b, 2013d).

Another example of such covenants and agreements comes from the Netherlands. Here the Dutch Government has entered into covenants with a range of individual businesses under the Green Deal programme, which sees the Dutch Government seeking to support local sustainability projects that face financial, regulatory or organizational barriers.

By 2011 over 150 Green Deal covenants had been signed between the Dutch Government and private sector actors, with a number of these seeking to improve urban sustainability (Ministry of Economic Affairs, 2013). At least 15 of these Green Deals seek to develop tools to give insights into the environmental performance of buildings. Some of these may help architects and engineers to meet requirements set in Dutch building regulation, while other tools help leaders in the industry to showcase that their buildings perform beyond requirements set in Dutch building regulation. The first of these tools were in operation by 2013, which is a remarkably short time of development. In one specific example 700 architects were involved in the development of a tool, and this broad involvement is considered to have resulted in overall support in the building design sector for this tool (OndernemendGroen, 2013).

The Dutch Government has financially supported the initiators of this Green Deal and provided administrative support in bringing a wide range of private sector organizations together. The strength of the Green Deals is the coordination and involvement of the Dutch Government in the various tools, test cases and best practices that are being developed and implemented. In particular, the role of an independent agency funded by the Dutch Government appears critical. This agency, the Netherlands Enterprise Agency, is responsible for administratively supporting the Green Deals, for bringing together relevant parties, and for drawing and communicating lessons about the Green Deals. This provides for a structured approach to drawing lessons as well as a central and easily accessible database for these lessons.

Partnerships and networks in Sydney

Typical examples of partnerships and networks are the Better Building Partnerships in London, Toronto and Sydney. These partnerships seek to bring together city councils with major commercial property owners, and, in the case of Toronto, with property developers and building users. In Sydney, for instance, the partnership brings together the city's 14 major property owners that together represent over 50 per cent of the office space area of the city's central business district. Commercial office space in Sydney produces about 50 per cent of the city's greenhouse gas emissions, which means that the partnership can, hypothetically, reduce up to 25 per cent of the city's emissions.

Under the partnership the council seeks to support property owners in terms of reducing the legal barriers and commercial risks of environmental upgrades of commercial property. In particular, it seeks to involve these property owners in policy-making processes so that the latter can plan their property portfolios accordingly. In return, the property owners make public commitments to reduce carbon emissions significantly beyond Australian regulatory requirements. Such reductions are important for the council as it has made an international pledge to reduce the city's overall carbon emissions by 70 per cent by 2030 (of 2006 emissions) and become one of the world's leading cities in terms of urban sustainability (City of Sydney, 2011).

The partnership was launched in 2011 and by the end of 2013 it had already reported considerable successes (Better Buildings Partnership, 2013): the total stock of the buildings owned by the 14 participating property owners had reduced its emissions by 7.5 per cent (of that in the starting year 2011), a doubling of the goal set, which reflects an emission reduction of 31 per cent of 2006 emissions. In addition, these buildings together had saved A$25 million annually from avoided electricity costs (Better Buildings Partnership, 2013). The engagement between regulators and the 14 property owners, and the city council's efforts to champion, promote and communicate the outcomes of the partnership to the wider community were considered the key factors of success by its administrators (Van der Heijden, 2013b, 2013d). The development of a database for tracking the performance of the building stock was considered one of the critical factors for the success of the partnership by its participants (Better Buildings Partnership, 2013).

On a different scale, CitySwitch Green Office seeks to improve the energy efficiency of Australian offices. CitySwitch is a network of Australian governments and office tenants. It was started by the City of Sydney in 2010, but due to its local success it quickly became a nationwide network. CitySwitch is administrated by local councils and state governments and serves as a platform for office tenants to learn about energy efficiency, share information, network and showcase good practices. The network aims to show tenants that there are limits to what they can do by changing work processes or making changes to a building's interior; and that more may be expected if their landlord makes significant changes to the building as a whole. As well as informing tenants, this network helps tenants to put pressure on their landlords to improve the environmental performance of their buildings. Individual tenants often have limited power in a tenant–landlord relationship; a collective of tenants has more power, and even more so when supported by a government agency.

By participating in the network, office tenants come to agreements with councils about their future environmental performance, and the council then provides support to help them meet these goals. In short, tenants agree to meet a particular rating within the voluntary NABERS benchmarking tool (further discussed in Chapter 4), which seeks to reduce energy and water consumption of existing buildings (NSW Government, 2011). By including this measurable goal, the risk of symbolic participation, or greenwash (Lyon and Maxwell, 2006), is partly overcome. Certain councils provide financial support to tenants, while others facilitate meetings and ensure an ongoing supply and distribution of information. In return for signing an agreement with a local council on future targets to be met, participants may use the promotional CitySwitch logo; and awards have been introduced to recognize leading practice.

The network's 2012 progress report claims that, compared to non-participants, CitySwitch participants 'boast an above NABERS rating' (CitySwitch, 2013, p. 14). This would indicate that participants on average use less energy and water in their buildings than non-participants. It remains unclear, however, whether CitySwitch attracts highly ambitious participants in the first place (that is, those who already have high NABERS ratings) or whether they become ambitious through participating.

As with many of this type of network, a highly accessible website provides a wealth of information to office tenants in the form of advisory reports on topics such as energy-efficient lighting, behaviour change and financing sustainable office upgrades, as well as a series of carefully documented best practices (CitySwitch, 2013). By the end of 2013 over 500 tenancies had committed to CitySwitch. The network's administrators consider the availability of non-cost information, the ability to showcase leadership and the nationwide recognition of the CitySwitch logo the key factors for its success (Van der Heijden, 2013b, 2013d).

3.2.3 Private Sector-led Collaborations

Mumbai's leading businesses

Where the Better Buildings Partnership and CitySwitch are examples of large collaborations that were initiated by governments, a typical example of such a collaborative initiated and led by private sector actors is Bombay First in Mumbai, India. This private sector to government collaboration was launched in 1994, building on a similar partnership in the United Kingdom, London First.

Bombay First was initiated as a solely private sector partnership. Its initiators aimed to make Mumbai a more liveable city, but also a city that would attract investment. To quote its chairman, 'Bombay First … aims to serve the city with the best that the private business can offer. [It aims to become] a magnet for the corporate world, a financial rival to top cities' (Nayar, 2010). The initiators of Bombay First expected that the liveability of the city, and especially for its urban poor, would significantly improve if the city attracted international investors. In 2003, together with the government of the state of Maharashtra, of which Mumbai is the capital, Bombay First commissioned McKinsey, an international management consulting firm, to prepare a study that assessed Mumbai's strengths and opportunities, in the process creating a vision for transforming the city. This vision document, *Vision Mumbai: Transforming Mumbai into a World-class City* (McKinsey, 2003), was endorsed by the government of the state of Maharashtra and the World Bank. With this endorsement the transition from a private partnership to a private–public collaborative network was made (Nayar, 2010).

Bombay First may be considered one of India's most successful examples of private sector to governance collaborations, which seeks to improve urban sustainability and resilience (PEARL, 2011). One of the major achievements of Bombay First is the development and implementation of the Mumbai Transformation Programme. This programme comprises more than 40 projects to improve economic growth in Mumbai, reduce poverty and enhance quality of life for residents, especially slum dwellers (Cities Alliance, 2008). The programme is financially supported by the JNNURM (discussed earlier in this chapter). Bombay First's successes are mainly attributed to the support of high level politicians, the commitment and involvement of businesses and especially its chairman, an organizational structure that can withstand political changes and seeking public awareness for the collaboration through the media (Cities Alliance, 2013).

A major player in ICT

Another example of a private sector-led collaborative network is the Connected Urban Development Programme. This collaboration brings together Cisco, an international developer of networking equipment, and the cities of Amsterdam, San Francisco, Seoul, Hamburg, Lisbon and Madrid.

Cisco was the initiating actor and started the programme as part of its commitment to the Clinton Climate Change Initiative, an initiative that builds on former US President Bill Clinton's commitment to the environment, and that implements programmes that create and advance solutions to the root causes of climate change. The network started in 2006 and was active for a period of five years. It sought to understand how innovative ICT applications can reduce greenhouse gas emissions in cities.

In 2010 the network was transferred into the SMART 2020: Cities and Regions Initiative. It found that with technologies available in 2008 about 15 per cent of global emissions could be saved by 2020 (Climate Group, 2008). This being due to, among others, reduced traffic in cities, more efficient use of energy in buildings and more efficient use of workspaces in buildings. In other words, most of the solutions identified by this network were related to the improvement of existing buildings and infrastructure.

The main lessons learned from this network do not so much relate to how to achieve more sustainable and resilient urban environments but how to achieve successful collaboration between city governments and the private sector (SMART 2020, 2013). Involvement of citizens in the programme was found to have resulted in public commitment and practical outcomes, and the setting of milestones was found to be key in keeping the appropriate level of focus between the organizations in the collaborative network. In addition, the involvement of visionary mayors who were already pushing a strong environmental agenda provided the leadership needed for the programme to achieve its outcomes (Boorsma and Wagener, 2007).

The Australian insurance industry

The Australian Resilience Taskforce seeks to encourage built environment resilience to extreme weather events. To reduce current risk to a tolerable level it considers that 'a focus on building standards, appropriate land-use planning and effective hazard mitigation' is necessary (Australian Resilience Taskforce, 2013). The taskforce was established by the Insurance Council of Australia, the representative body of the general insurance industry in Australia. It is the council's response to a report on climate change by the Australian Government that considers the cost of compulsory insurance as a potential incentive for citizens and organizations to seek higher levels of resilience of their property (Productivity Commission, 2012). The taskforce is a platform for collaboration across government, industry and NGOs (Australian Resilience Taskforce, 2012).

In order to achieve a more resilient built environment in Australia, the taskforce has developed a Building Resilience Rating Tool in alliance with 120 stakeholders from various backgrounds in the building sector. The tool builds on a database of the resilience of building materials and products to extreme weather events, and a database on natural hazards and their probability of occurrence. The first database is compiled and administered by the taskforce. The Building Resilience Rating Tool allows for an assessment of a building's risk to particular natural hazards. The current tool addresses hazards classified as 'inundation' and 'storm'; ultimately the tool will also address hazards classified as 'cyclone', 'bushfire', 'earthquake' and 'extreme heat'. These are all risks that are not addressed as such in the Building Code of Australia.

The key factor for success, the rapid development of the tool, appears to be the well-organized Australian insurance industry that was able to activate a large number of participants to develop the Building Resilience Rating Tool, and to seek attention to the problem of lacking resilience of the Australian built environment.

On an international level the ClimateWise group, a network of the global insurance industry's leaders, and the Institutional Investors Group on Climate Change, a collaboration between international pension funds and other institutional investors, are seeking to address issues comparable to the Australian Urban Resilience Taskforce (ClimateWise, 2012; IIGCC, 2013).

It goes without saying that for all three examples of private sector-led collaborations another factor was in place that may have aided their successes. If the collaborations are successful in seeing their ideas taken up by policy makers, they are almost certain of opening up new markets. The participants in Bombay First will be at the forefront to welcome international investors as well as government funds, Cisco has a clear private interest in seeing its ICT networking solutions being implemented by different cities in the world, and the participants in the Australian Urban Resilience Taskforce may see their insurance claim payouts go down and their insurance incomes go up if insurance policies in Australia are being changed. In these examples the line between collaboration and lobbying may be thin.

3.2.4 Civil Society-led Collaboration

Major NGOs and non-profits

Civil society-led partnerships and networks partly echo the structure of earlier discussed examples, especially when led by large and well-organized NGOs or non-profits. For instance, in 2013 the Rockefeller

Foundation initiated the 100 Resilient Cities Centennial Challenge. This challenge seeks to establish a network of 100 cities that will seek to improve their responses to the increased resilience problems of ongoing urbanization, population growth and climate change. The first of the 100 participating cities were selected out of 400 that applied for participation in the challenge. Selected cities are financially supported by the Rockefeller Foundation by hiring a well trained individual to oversee the development of a resilience strategy of the city and to communicate with other cities in the challenge. They are administratively supported in the development of a resilience plan. Finally, they find support from other members in the network and actively share lessons and best practices on how to achieve urban resilience (Rockefeller Foundation, 2013). The challenge shows remarkable comparisons with the C40 Cities Climate Leadership Group and ICLEI discussed earlier.

Another typical example of a collaboration led by a large and well-organized civil society organization is the Common Carbon Metric (UNEP, 2010), launched by the United Nations Sustainable Buildings and Climate Initiative at the Climate Change Conference in Copenhagen (COP 15) in 2009. This tool seeks to address the global need for a unified approach to the reporting and measuring of carbon emissions from the global built environment. It is expected that the metric will lead to the establishment of benchmarks and baselines to make meaningful comparisons between buildings, and to set verifiable greenhouse gas reduction targets for the built environment. This may then provide a consistent basis for future monetization of carbon trading measures in the building sector (Atkinson, 2010).

For such an ambitious project, the United Nations relatively quickly developed the metric, within less than a year, and had the metric ready for pilot studies by 2010 (UNEP, 2010). The development and piloting processes were let by the United Nations, and governments, NGOs and businesses worked together in these processes. Although the pilot phase of the metric appears to have ended, there does not seem to be wide-scale implementation of the tool.

Early in 2012 the International Standards Organization (ISO) made reference that it was considering a draft standard based on the metric (ISO, 2012b), yet not much information is currently (2014) available on the status of this tool. It is telling that by the end of 2013 the discussion group dedicated to the Common Carbon Metric on LinkedIn, a social networking website for people in professional occupations, only had six participants (LinkedIn, 2013).

The Center for Neighbourhood Technology

A somewhat different example is the Sustainable Backyard Program in Chicago, Illinois. This is a partnership between the city of Chicago and the Center for Neighbourhood Technology, a non-profit charitable organization. It was originally initiated by the city of Chicago but is now led by the Center for Neighbourhood Technology. It is an educational and incentive programme that helps Chicago homeowners to create more sustainable landscapes, starting in their own backyard. The programme was created to reduce storm runoff water into Lake Michigan and other waterways in Chicago, through green infrastructure. By using green infrastructure, such as gardens, rainwater can be naturally filtered, stored and reused. At the same time, green infrastructure may help prevent flash floods that could originate from water running down hard surfaces – that is, streets, concrete slabs and pavements.

The programme encourages homeowners to conserve water and manage storm water. It seeks to educate homeowners about backyard composting, harvesting rainwater, making a rain garden, planting trees and other activities that will make Chicago more sustainable and resilient. Through such education and activities it is hoped that participants will improve their environmental performance in other parts of their lives as well.

The programme's main incentive is a rebate of up to 50 per cent for homeowners if they buy trees, native plants, shrubs, compost bins and rain barrels from participating suppliers. In addition, workshops are organized to inform homeowners on how they can significantly improve their environmental performance, even through small changes in their lifestyle or behaviour. It remains unclear whether the programme has achieved successful outcomes (Van der Heijden, forthcoming 2014a). One of the initial barriers faced by its administrators was outreach. It appears difficult to ensure that the programme is widely known by homeowners and retailers alike (Chicago Garden, 2012).

A group of homeowners

On a different scale again, Green Strata is a group of active homeowners. Initially, it sought to get legislation changed that hampered the instalment of solar photovoltaic systems on multi-unit residential buildings in Australia. Green Strata was started by a number of owners in Sydney in 2010, but has quickly grown to become a national programme. The initial participants successfully connected with the Sydney City Council, who provided Green Strata with a small fund to cover its start-up costs.

The fund was used to develop a website that provides information to owners of units in multi-residential buildings and their managers on how to improve the environmental sustainability of these buildings (Green Strata, 2013). This website is an easily accessible source of data with case studies and video interviews that give insight into particular approaches to improve the environmental sustainability of strata buildings, and with information from suppliers that may help strata owners to improve the sustainability of their buildings.

In collaboration with the Sydney City Council, Green Strata has been a driving force in the development of the Smart Green Apartments programme that was launched in 2012. The programme aims to help create more cost-effective and resource-efficient buildings, improve performance of shared services and amenities and minimize environmental impacts. The focus of the programme is to improve the energy efficiency of the shared spaces of strata buildings. In the first two years of the project 30 pilot buildings were engaged. Important lessons were drawn on how relatively easy significant reductions in energy consumption could be achieved. For instance, in one of the pilot buildings 82 per cent of its power use was saved by merely upgrading fire escape lighting (City of Sydney, 2013).

Green Strata was further involved in the Smart Blocks programme, a nationwide Australian private sector to government collaboration between strata owners, the cities of Sydney and Melbourne and the Australian Government. This programme seeks to engage with strata owners in saving energy in their apartment blocks. The original initiators of Green Strata consider a number of factors to have added to the successes of the collaborative (Van der Heijden, 2013b, 2013d). The initiators all had extensive business experience as senior managers in various positions, which they could build on when developing Green Strata. This helped them in applying for grants and in building strong relationships with the city of Sydney. In addition, they considered the idea of 'owners helping owners' as a key factor for success. This may give the collaboration a high level of credibility for prospective participants.

An architect with a mission

Open Mumbai seeks to improve the environmental sustainability of Mumbai and to improve the liveability of the city, and in particular to improve the living conditions of the urban poor (Thirani, 2012). Open Mumbai is a result of over 15 years of activism and civil society to government collaboration led by a Mumbai-based architect, PK Das (Das, 2012). Open Mumbai looks beyond the land officially earmarked as open space for providing urban sustainability – that is, recreation grounds, playgrounds, gardens, parks and hills. It considers the city's mangroves, rivers and *nullahs* (streams) as relevant areas. Currently, however, many of these are highly polluted.

Through creating eco-sensitive borders along mangroves, for instance, Open Mumbai seeks to combine public open space with urban sustainability, while at the same time preserving and revitalizing the natural assets of the city. By creating such recreational eco-sensitive borders open public spaces can be connected. As the architect states, 'the idea of creating green spaces is not just designated to the building of cute and fancy parks and gardens but creating a network of open spaces, open and clear forever for all the citizens equally' (Das, 2013).

The participants in Open Mumbai have much experience with involving local citizens (slum dwellers and others) in such projects, and through their project they strive for affordable housing throughout the city. Since the late 1990s they have revitalized a number of Mumbai's beaches and coastlines, and some of its parks in projects that brought together citizens and local governments. They consider such citizen involvement as critical for the success of the revitalization of Mumbai (Carr, 2013). Ideas underpinning Open Mumbai have been formalized into the city's official development plan, which may very well be the biggest achievement of its participants (Thirani, 2012). Notwithstanding the important role of citizen participation in the various projects that have cumulated in Open Mumbai, it appears that its major reason for success is the leadership role of the aforementioned architect.

3.3 CONSIDERATIONS

The real-world examples discussed largely confirm the extant literature on collaborative governance discussed in the first section of this chapter. Collaborative governance can be considered a highly pragmatic approach to public problem solving: it seeks for solutions that are suitable to a certain context in a certain time (Karkkainen, 2004; Solomon, 2008; Walker and de Búrca, 2007).

The majority of the collaborative governance tools discussed in this chapter are characterized by a willingness to learn and share lessons, or even have made learning and sharing lessons one of their core activities, such as ICLEI and CitySwitch Green Office. This focus on learning, sharing lessons and adjustment to lessons learned is a major difference between the tools discussed in this chapter and the more traditional governance tools discussed in Chapter 2 (Ford, 2008; Hertier, 2002; NeJaime, 2009). It is, however, not to say that collaborative governance is a panacea for the shortfalls of direct regulatory interventions.

3.3.1 Some Concerns About Collaborative Governance for Urban Sustainability and Resilience

Perhaps the most fundamental concerns related to collaborative governance are its two major implementation difficulties. In real-world settings it often appears difficult to achieve a true sharing of power among the participants in collaborative processes; and because of the large number of participants involved, collaborative processes may face collective action problems.

In other words: What roles do participants have in collaborative processes? What is exactly asked of them? What exactly is needed of them to develop governance tools that achieve desired outcomes? Should all relevant stakeholders have an equal level of decision-making power? Should all participants agree with the outcome of the process? It goes without saying that answering these questions with 'yes' would result in unworkable governance processes.

Although there is no clear set of guidelines that will guarantee successful outcomes of collaborative processes, it is exactly in answering such questions, before starting a collaborative governance process, that expectations of potential participants can be managed. Such clearness about the process may prevent participants from feeling that they are not taken seriously, which may result in them not trusting or accepting the outcome of the process (Irvin and Stansbury, 2004; Lukensmeyer and Torres, 2006).

These practical difficulties of collaborative governance processes have been recognized for some time (for example, Arnstein, 1969; Collins and Ison, 2009; Irvin and Stansbury, 2004; Konisky and Beierle, 2001; Olson, 1965; Stirling, 2004; Webler and Tuler, 2006). Despite these complications, collaborative governance is still being widely pursued and preferred by governments, businesses and civil society groups and individuals as an approach to governance that may bring promising outcomes – this was stressed by the large number of examples discussed in this chapter.

It could be argued that collaborative governance is both an ideology of how governance tools ought to be developed and implemented, as well as a highly pragmatic process to develop these tools. As an ideology it comes with many expectations about increased effectiveness, efficiency, accountability and legitimacy of governance tools. These expectations have, however, not yet been unequivocally supported by the research into collaborative governance carried out thus far (Karkkainen, 2004; Schout et al., 2010). The lack of support for these expectations is, in part, exactly related to the practical difficulties of collaborative governance.

The examples discussed provide two clear additional illustrations of such practical difficulties.

The example of the Brisbane community plan highlights that a government-driven collaborative approach may be captured by policy makers or administrators. It is striking to see that in two collaborative processes the citizens of Brisbane have made such a major swing in their opinion. In 2009 they wanted their city to become Australia's most sustainable city and have its council carbon neutral by 2026, while just four years later, in 2013, they had left these ambitions and were more interested in the economic growth of their city. These are two remarkably different outcomes of two collaborative processes in which the same citizens (may) have been involved.

Although the two examples do not give direct evidence that the architects of the collaborative processes that led to the Brisbane community plan in 2009 and the Brisbane City plan in 2013 intentionally designed these processes to achieve a particular outcome, it is difficult to believe that the same citizens have come to such different opinions. These different outcomes are more likely explained by differences in the design of the participatory processes, the involvement of different groups of citizens in the processes and different political preferences by the architects of the collaborative processes (Arnstein, 1969; Lindsay and McQuaid, 2009; Pennington et al., 2011).

Another additional illustration of the practical difficulties of collaborations is that they run the risk of becoming elite networks that exclude non-members from lessons learned and other advantages (see also Kern and Alber, 2010). The 100 Resilient Cities Centennial Challenge, the Better Building Partnership, the C40 Cities Climate Leadership Group, CitySwitch Green Office and the SMART 2020: Cities and Regions Initiative all present some best practices and case studies on their websites, but they all have members-only sections that provide more information, or provides information well before it is made public.

There is, of course, a good reason why these collaborations exist. Participation is made attractive by rewarding participants with some privileges unavailable to non-participants. In other words, why would participants care to join a collaboration when doing so does not make them better off than non-participants? However, particularly when governments are involved in the development, implementation and financial support for collaborations, the rewarding of participants over non-participants is questionable. Taxpayers' money is then likely used to support these collaborations, and it is questionable whether governments can legitimately favour participants over non-participants in (indirectly) gaining from such support. This argument aligns with the oft-made argument that subsidies and tax rebates may strengthen inequalities, especially because they often transfer general taxes to the already well off in societies (see Chapter 2).

Governmental support in collaborations may also result in situations where those participating in collaborative governance processes only do so because they get or expect to get financial or other forms of rewards from participation. In such circumstances participants may act in a way they think they are expected to act in order to secure these rewards, which may endanger the long-term viability of the tools developed in collaborative processes (Backstrand et al., 2010; Héritier and Eckert, 2008; Shimkada et al., 2008).

Finally, there is no solid evidence that collaborations outperform direct regulatory interventions in addressing urban sustainability and resilience (see Van der Heijden, forthcoming 2014a; Wurzel et al., 2013; Young et al., 2008). The wide range of examples discussed points to much activity in this area, but the examples themselves overall do not show sweeping outcomes. In some contexts meaningful results are achieved (for example, CitySwitch Green Office), while in other contexts this is not the case (that is, the Common Carbon Metric). In other examples different participants achieve different results. The length of membership in ICLEI, for example, appears related to participants' performance. The longer a city is a member of ICLEI, the better its environmental performance (Reams et al., 2012).

The main lesson for governments, businesses and civil society groups and individuals interested in this approach to governance is that collaborative governance should be approached with some care. Ideological aspirations may easily stand in the way of the practical implementation of collaborative governance and may blind participants to negative outcomes (Termeer, 2009).

3.3.2 Potential to Overcome the Three Main Governance Problems

Whether and how is collaborative governance able to overcome the three main governance problems discussed in the introduction to this book? Some of the examples discussed directly target one or more of problems related to direct regulation:

- The Better Building Partnership and Green Strata, for example, have proved successful in addressing the problem of grandfathering. In the Better Building Partnership commercial property owners have, supported by the City of Sydney, achieved considerable successes in reducing the energy consumptions of their existing buildings.

- Green Strata highlights that highly active, capable and maybe even persistent individuals may very well make the difference between highly successful outcomes of partnerships and networks and less successful ones. Here governments may seek to take up new roles to support and empower such individuals in the development of partnerships and networks, and to connect them with others. Also, the examples discussed indicate that much activity can be uncovered in terms of partnerships and networks that address urban sustainability, but it remains unclear whether urban resilience is addressed to the same extent by such collaborations. The current research has not uncovered many examples of these, which is, of course, not to say there are none.
- Examples such as PEARL, Open Mumbai and Bombay First illustrate that, at least in India, the time lapse between the development and transformation in rapidly developing economies and their governments' regulatory responses can be addressed through partnerships and networks.

It is, however, questionable whether such examples can be replicated in other contexts. Three individuals, the chairman of Bombay First, one of the initiating administrators of PEARL and the architect in Open Mumbai, appear to have fulfilled pivotal roles in these collaborations. Specific individuals also played a strong role in making Green Strata a success. These individuals' enthusiasm, vision and drive may very well have made the difference in these examples being fairly successful. Again, local governments may wish to take up supportive, empowering and assembling roles to ensure that such individuals can flourish in the development of collaborations. International governments, businesses, NGOs and civil society groups may seek to do the same.

Examples such as CitySwitch Green Office, Green Deals and the Climate Change Sector Agreements give some insight as to how collaborations may be successful in addressing parts of the wicket set of market barriers discussed in Chapter 1:

- The Green Deals and Climate Change Sector Agreements stimulate leadership in the industry, reward such leadership and showcase such leadership. This may partly take away the first-mover barriers developers and property owners may experience.
- CitySwitch seeks to involve building tenants in a move towards improved urban sustainability by seeking to change their behaviour.

> Through CitySwitch tenants may also be empowered to successfully engage with their landlord about retrofits of the buildings they rent, which addresses the split incentive barrier these actors face.

These are hopeful examples, but at the same time they appear to be pockets of good practice in a further highly unsustainable sector (Van der Heijden, forthcoming 2014a, 2014b, 2014c). It is unlikely that this wicket set of market barriers can be addressed one by one. Governments may again wish to take up supportive and in particular assembling roles to link various collaborations.

The Sydney City government appears to have well understood the importance of assembling these different collaborations (and, as highlighted in Chapter 4, assembling these with voluntary programmes and market-driven governance tools). By being involved in a number of collaborations it can link property owners to property tenants (that is, participants in the Better Buildings Partnership to participants in CitySwitch), which may result in synergies between the various collaborations. The Netherlands Enterprise Agency provided another example of how governments can take up initiating and assembling roles in collaborative governance.

3.3.3 Promises for Governing Urban Sustainability and Resilience

In conclusion, the shift towards collaborative governance is a promising turn for achieving urban sustainability and resilience.

Cities are the level at which many problems related to global changes originate and manifest themselves – that is, unsustainable levels of resource consumption, rapid economic development and urbanization in the global South, a growing world population and increased climate risks. It thus makes sense to address problems where they originate, and use local knowledge to address problems (Borghi and Van Berkel, 2007; Lobel, 2004; Noveck, 2011; UNCED, 1992).

The drawbacks of this approach to governing urban sustainability and resilience relates to the danger of capture of collaborative processes by strong stakeholders at the expense of weaker stakeholders, and lack of power to enforce the fulfilment of urban sustainability and resilience goals. Besides, the results of collaborative governance tools appear highly context and participant dependent. Care should be taken when generalizing from a relatively small set of experiences (as in the current chapter), and even more care should be taken by governments, businesses and civil society groups and individuals when seeking to copy a successful collaboration from elsewhere to their own context. Successes elsewhere are by no means a guarantee for successes at home.

4. Voluntary programmes and market-driven governance

It is a little step from collaborative governance to voluntary programmes and market-driven governance, and the exact dividing line between these approaches to governance is difficult to draw. Both build on voluntary participation in governance tools combined with clear financial or other direct rewards for their participants. This makes these tools somewhat different from the collaborations discussed in the previous chapter, which often build on the sharing of lessons and information about how to achieve urban sustainability and resilience.

At first glance the term 'voluntary programme' seems to imply that this approach to governance is fully non-coercive, while the term 'market-driven governance' seems to imply that this approach to governance is without government involvement, and thus voluntary to a certain extent as well. Yet, studies on both approaches point to a more complicated picture.

Voluntary programmes often echo the structure of direct regulatory interventions (Potoski and Prakash, 2009), with a set of rules that its participants are expected to follow. Often these rules are enforced, at least to a certain extent; and non-compliance may in certain programmes lead to disciplinary action. Market-driven governance, in turn, is often found to have some form of government involvement (Cashore et al., 2004).

Voluntary programmes and market-driven governance as approaches to governance have become popular since the 1990s. Like collaborative governance, they fit the shift 'from government to governance' discussed in Chapter 3. It can be argued that the current uptake and interest in these follows on from an earlier interest in self-regulation (Andrews, 1998; Gunningham and Grabosky, 1998; Sinclair, 1997).

Self-regulation refers to situations where businesses or citizen groups regulate their own behaviour. Sometimes they do so under the threat of direct regulatory interventions. It is then expected that those targeted by self-regulation will implement regulatory interventions that are more cost-effective than the interventions government may introduce, while still achieving the desired collective end (Konar and Cohen, 2001;

Saurwein, 2011). Government can 'threaten' to implement statutory regulation unless businesses or citizens take action themselves.

An example is the Climate Wise programme in the United States. This programme sought to incentivize non-utilities to reduce their greenhouse gas emissions at a time when there was a significant threat of the introduction of direct regulatory requirements. Climate Wise was established in 1993 and remained in force until 2000. The programme, however, only achieved limited results, partly because the threat of direct regulatory interventions was too weak. The targeted industry did not experience sufficient pressure to actually change their behaviour (Morgenstern and Pizer, 2007; Pizer et al., 2010). Examples like these, and others (Saurwein, 2011), have meant that self-regulation is now looked upon with some suspicion, especially as an approach to environmental governance (Potoski and Prakash, 2005b).

Interestingly, however, voluntary programmes and market-driven governance seek to achieve behavioural goals through highly comparable mechanisms as self-regulation. With their increasing popularity, the question then becomes: To what extent are they promising approaches to governance for urban sustainability and resilience?

4.1 CHARACTERISTICS OF VOLUNTARY PROGRAMMES AND MARKET-DRIVEN GOVERNANCE

The literature regarding voluntary programmes and market-driven governance shows much overlap with the literature on collaboration as an approach to governance (Van der Heijden, 2012, 2013c). This has to do, in part, with the fact that voluntary and market-driven interventions are often developed by businesses or citizen groups in collaboration with governments. The advantages and disadvantages of such collaborative development processes have been discussed to some extent in Chapter 3. To prevent too much overlap with Chapter 3, the discussion that follows therefore addresses in particular the development and implementation process and structure of the voluntary programmes and market-driven governance tools, and the motivations for non-governmental actors to voluntarily subject themselves to these tools.

4.1.1 Actors Involved

The existing literature distinguishes three main reasons why governmental and non-governmental actors are involved in the development and implementation of voluntary programmes and market-driven governance tools.

First, sometimes governments solely develop these tools (Darnall and Carmin, 2005). They then may seek to nudge or incentivize non-governmental actors to voluntarily move their behaviour beyond the requirements set by legislation or regulation. The difficulty of implementing future regulation, or difficulties of compliance with existing regulation, may provide another reason for governments to seek voluntary programmes and market-driven governance tools as alternative governance approaches to direct regulatory interventions (Delmas and Terlaak, 2001; Lyon and Maxwell, 2007; Trubek and Trubek, 2007; Weber and Hemmelskamp, 2005). Government may also use its potential as a major customer to seek a voluntarily change in its suppliers' behaviour. By demanding, for example, highly sustainable buildings it may act as 'launching customer' for these voluntary programmes and market-driven governance tools (Hofman and De Bruijn, 2010; Van der Horst and Vergragt, 2006).

Second, the literature points to a series of voluntary programmes and market-driven governance tools that are the result of collaboration between governments and non-governmental actors (Ansell and Gash, 2008; De Clercq, 2002; Freeman, 1997; Mol et al., 2000). Examples, such as negotiated agreements and covenants, have already been discussed in Chapter 3. Other examples are the immensely popular best-of-class building benchmarking tools such as Leadership in Energy and Environmental Design (LEED) that are often developed by non-governmental actors in collaboration with governments (Lee and Burnett, 2008; Schindler, 2010) – these tools are further discussed in Section 4.2.

Third, non-governmental actors may seek to develop such tools without any governmental involvement (Cashore et al., 2004; Cooper and Symes, 2009). Again, best-of-class benchmarking tools stand out as typical examples. While the involvement of non-governmental actors in these approaches to governance seems obvious, it brings to the fore questions of how these tools interact with existing legislation and regulation, and whether and how these tools can overcome the governance problems discussed in Chapter 1 (Cafaggi and Janczuk, 2010; De Bruijn and Norberg-Bohm, 2005; Trubek and Trubek, 2007). In addition, existing legislation or regulation may support or hamper the development and implementation of voluntary programmes and market-driven governance tools, which may explain why this approach to governance is more popular in some countries than others (Croci, 2005; EEA, 1997).

4.1.2 Development and Implementation: Why Would Participants Voluntarily Submit Themselves to a Governance Tool?

In this context the word 'voluntary' is potentially confusing, and requires some level of explanation (Darnall and Sides, 2008; Lyon and Maxwell, 2007; Potoski and Prakash, 2009; Segerson and Miceli, 1998).

When exactly can we say that an individual or an organization does submit itself voluntarily to a governance tool? Is it the case when there is no governmental involvement at all? Is it the case when doing so does not serve this individual or organization's private interest? Is it the case when there is no societal pressure at play?

Time and again, the literature indicates that often individuals and organizations do submit themselves voluntarily to a governance tool simply because they are subject to internal and external pressures (Baranzini and Thalmann, 2004; Potoski and Prakash, 2009).

The impact of future regulation is often highlighted as a major reason why non-governmental actors develop and participate in voluntary programmes (Maxwell et al., 2000; Reid and Toffel, 2009). It is expected that they do so to prevent or stall the implementation of future regulation. After all, it is likely that regulations they develop themselves suit their interest better than regulations developed by governments. Alternatively, non-governmental actors may seek to steer governments to introduce particular regulations that makes it more difficult for new business to enter the market (Barrett, 1991; Salop and Scheffman, 1991).

Another reason why non-governmental actors develop and participate in voluntary programmes and market-driven governance is found in market conditions. From a purely rational point of view the only logical reason for them to comply is if the profit of development and participation outweighs the costs (Croci, 2005; Howarth et al., 2000).

A more demand-driven motivation is found in the literature on 'green consumerism' and 'green investors'. Green consumerism implies consumers' demands for products that have better environmental performance than that required by governmental regulation, and a willingness to pay a premium for such products (Arora and Gangopadhyay, 1995; Baron and Diermeier, 2007; Feddersen and Gilligan, 2001). Green investors implies investors' concerns with a firm's environmental performance (Graff Zivin and Small, 2005; Hamilton, 1995; Hoffman, 2001), which 'could be viewed as exposing the firm to greater risks of liabilities, penalties and high costs of compliance in the future, or reflecting poor management practices and lack of innovativeness' (Khanna and Anton, 2002, p. 543).

Societal pressure is considered a third reason why non-governmental actors voluntarily develop and participate in this type of governance tools (Briscoe and Safford, 2008; Den Hond and De Bakker, 2007; Mikler, 2009). The literature is aware of the role of NGOs in directly influencing firms and businesses, for instance, by addressing the board of a large firm or by mobilizing on a certain site to prevent a firm's operation (Baron and Diermeier, 2007; Bartley, 2003; King, 2008). Alternatively, NGOs may seek to impact firms and businesses indirectly, for instance, by influencing consumer behaviour through boycotts and negative information campaigns (Arora and Cason, 1995; Bartley, 2003).

4.1.3 Governance Tools: What is Their Structure?

The structure of voluntary programmes and market-based governance tools are highly comparable with that of direct regulatory interventions (Van der Heijden, 2012): rules; monitoring and enforcement; and sanctions. The literature is further aware of the importance of rewarding schemes in this approach to governance (Potoski and Prakash, 2009).

Generally, the bases of voluntary programmes and market-based governance tools are considered to be their rules or rule structures. Leading authors in the field, Matthew Potoski and Aseem Prakash, refer to these as 'club standards' (Potoski and Prakash, 2009, pp. 24–6). Others use terms such as 'value and goal statements' and 'plans or targets' (Darnall and Carmin, 2005, p. 77). These rules are implemented to prescribe the goals of the voluntary programmes and market-based governance tools, their expected outcomes and the behaviour that is expected from participants to achieve these outcomes. The discussions on statutory regulation in Chapter 2 provide an overview of the types of rules that are also used in voluntary programmes and market-based governance tools.

The literature on this approach to governance is highly aware of the need to monitor and enforce the performance of voluntary programmes and market-driven governance tools. Time and again, scholars find that without monitoring or enforcement participants do not conform to the rules (Bailey, 2008; Darnall and Sides, 2008; Delmas and Keller, 2005; Lyon and Maxwell, 2000; Rivera and de Leon, 2004). Different types of monitoring or enforcement are discussed: self-monitoring; administrator monitoring; monitoring by a third party hired by the participant; independent third-party monitoring; government monitoring.

It is generally considered that self-monitoring by participants, and monitoring by the administrators of these tools, are of little avail when aiming to bring participants into compliance (Darnall and Carmin, 2005; King and Lenox, 2000). More is to be expected from third-party monitoring such as certification or audits (Cashore et al., 2004; Darnall and Sides, 2008; Lyon and Maxwell, 2000) or governmental monitoring (Bartle and Vass, 2007; DeMarzo et al., 2005). Furthermore, external monitoring is easily recognized by stakeholders and may be considered as more legitimate than self-monitoring or administrator monitoring (OECD, 2003).

The capstone of monitoring and enforcement are the disciplinary measures taken when non-compliance is traced or rewards are given for compliant behaviour. The literature presents two lines of reasoning. Following common-good theorizing (Jordan, 1989), a strand of literature argues that the actual rewards of participating in a voluntary programme or market-based governance tool should compel participants to comply (King and Lenox, 2000). Yet, following deterrence theorizing (Hawkins, 1984; Reiss, 1984), another strand of the literature argues that more severe disciplinary measures are needed and the potential penalties arising from detected violation should be feared (Lenox and Nash, 2003). Different forms of penalizing are considered: financial penalties; withdrawal of a participant's membership; or publication of the names of those in violation (King and Lenox, 2000; Short and Toffel, 2010). Rewards for participants come in many forms:

- information (Lyon and Maxwell, 2007)
- access to public officials (Bischop and Davis, 2002; Hunold, 2001; Magagna, 1988)
- public recognition (Arora and Cason, 1995; Bansal and Hunter, 2003)
- financial profit (Croci, 2005; Howarth et al., 2000).

It is in these rewards that the motivations for developing and implementing this type of governance tools are confirmed.

4.2 FROM THEORY TO PRACTICE: URBAN EXPERIENCES WITH VOLUNTARY PROGRAMMES AND MARKET-DRIVE GOVERNANCE

In seeking to improve urban sustainability and resilience, businesses, NGOs, citizen groups and governments alike are highly active in these approaches to governance. There is no shortage of governance tools that build on voluntary participation, combined with clear financial rewards for their participants.

In what follows, a series of examples is addressed. For heuristic purposes these are clustered as classification and best-of-class benchmarking tools (Subsection 4.2.1), green leasing (Subsection 4.2.2), private regulation (Subsection 4.2.3), innovative financing (Subsection 4.2.4), contests, challenges and competitive grants (Subsection 4.2.5), intensive behavioural interventions (Subsection 4.2.6) and sustainable procurement (Subsection 4.2.7).

4.2.1 Classification and Best-of-Class Benchmarking

In 1990 the UK-based Building Research Establishment (BRE) wrote history by certifying the world's first BRE Environmental Assessment Method (BREEAM) building. This certification evidences the building's leadership in environmental sustainability: it is benchmarked as the best of its class. Best-of-class benchmarking fits a global trend of building classification. The idea underlying best-of-class benchmarking is simple and elegant: by ranking a building in a certain class its performance in terms of energy, water and material use can easily be compared to that of other buildings of the same class – at least in theory. It is this ease in comparing that makes these benchmarking tools so attractive.

Best-of-class benchmarking tools show striking similarities with statutory building regulation and energy performance certification discussed in Chapter 2. In order to be certified, a building plan or construction work is assessed against a series of predefined regulations. Credits are awarded for each regulatory requirement met, and the more credits achieved, the higher the classification of the building. For developers, investors, property owners and occupants alike it is easy to understand that on a scale from poor performing to high performing, say one to five stars or bronze to gold, a five-star or gold-classed building is somehow better than a one-star or bronze-classed building (the words 'benchmarking', 'rating' and 'labelling' are often used interchangeably in this context; though they refer to slightly different approaches to classification; see Pérez-Lombard et al., 2009).

But not only are these benchmarking tools of interest to developers, investors, property owners and occupants. Governments around the globe participate in, support and sometimes even develop these tools as they generally seek a voluntary move beyond governmental construction codes and regulation for sustainability or in some countries the tools provide the only form of sustainable building regulation.

Since the early 1990s a range of comparable building classification tools have been implemented around the world, with some authors making estimations of at least 90 tools in place (Fowler and Rauch, 2006a). Best-of-class benchmarking, and related classification tools, may be considered one of the most popular voluntary market-driven governance tools in the building sector (Fowler and Rauch, 2006b).

Leading private sector benchmarking tools

In its 20 years of existence BREEAM has been marketed throughout the world and is now adopted in more than 50 countries. Over 15 000 construction projects have been certified, which equates to more than 200 000 buildings globally (BREEAM, 2013). In 1993 the US Green Building Council certified the first LEED building. As with BREEAM, LEED has been exported around the world and is now adopted in 135 countries and territories. Around the globe close to 20 000 projects have been LEED certified since then (USGBC, 2013b). This makes LEED and BREEAM some of the most widespread internationally applied market-driven governance tools, and the tools have been well reviewed in the academic literature (for example, Fowler and Rauch, 2006b; Horvat and Fazio, 2005; Lee and Yik, 2004).

Tools such as LEED and BREEAM come with accolades and critique alike. Some early research points out that there is an emerging market for benchmarked office space. The demand for sustainable office space appears partly related to a wish of organizations to showcase their 'sustainable' credentials (Dixon et al., 2009). The benchmark of their buildings, then, allows for a clearly visible and internationally accepted approach for showcasing these credentials.

Also, empirical research shows that sustainable office space may yield higher rents and higher selling prices (Eichholtz et al., 2010; GBCA, 2013a). However, the same research indicates that other factors such as location and building quality remain major drivers for occupants to rent sustainable office space.

Finally, by the end of the first decade of the twenty-first century, the costs of constructing high-performing buildings, which achieve high classifications in benchmarking tools, had become at par with those of constructing conventional buildings (Hoffman and Henn, 2009). Nevertheless, by then the majority of the building sector was still of the opinion that benchmarked buildings were more expensive to construct than conventional buildings (WGBC, 2013).

The successes of LEED buildings, in terms of energy reductions as reported by the US Green Building Council, has been questioned (Gifford, 2009). The tool was further critiqued for having a focus on assumed energy performance and not on evidenced energy performance. The initial approach of benchmarking tools, such as LEED and BREEAM, was to certify a building based on an assessment of its design (certified 'as designed') or based on a series of audits carried out during its construction (certified 'as constructed'). The true performance of buildings, however, will only become clear when they are in use.

Both LEED and BREEAM, as well as other benchmarking tools, have now introduced a category to assess buildings 'in operation' in order to be able to certify these on their actual performance (BRE, 2013a; USGBC, 2013c). This new category appears all the more important since these tools' actual performances continue to be questioned. For instance, there does not appear to be a correlation between energy savings of a LEED certified building and the number of credits the building was awarded (Newsham et al., 2009). Further, studies have indicated that LEED certified buildings do not outperform conventional buildings in terms of energy usage and greenhouse gas emissions (Scofield, 2009), and in certain examples they even seem to perform worse (Scofield, 2013).

Another oft-heard critique is that the tools allow for gaming (Hoffman and Henn, 2009). Some of the regulatory requirements these tools set are easier or cheaper to meet than others. The introduction to an article on a sustainable construction information website is telling:

> *How to Cheat at LEED for Homes*: The road to green certification is paved with low-hanging fruit. This cheat sheet with 22 shortcuts will get you to LEED certification without a lot of trouble. (Seville, 2011)

These '22 shortcuts' allow for 70 LEED credits, which is sufficient to have a building 'Gold' certified, the second highest tier of certification. LEED is also critiqued for not addressing the context of LEED certified buildings, or a more holistic approach to urban sustainability. How can, for instance, a parking garage be certified under the LEED tool, critics wonder (Alter, 2008).

An interesting example of a more recent best-of-class benchmarking tool, which explicitly rewards designers and developers for thinking more holistically about urban sustainability, is the Living Building Challenge administered by the International Living Future Institute. This building certification tool claims to be 'the built environment's most rigorous and ambitious performance standard' in the world (International Living Future Institute, 2014). The tool is of interest because of the synergies it seeks between various aspects of urban sustainability, and because it only awards buildings for their actual performance after at least 12 months of operation, and not their modelled or designed performance.

Other private sector and non-governmental classification and benchmarking tools

Not only have these two leading benchmarking tools been exported to many countries and regions, they have inspired others to develop their own benchmarking tools. For instance:

- Green Star in Australia
- the DGNB system (Deutsche Gütesiegel Nachhaltiges Bauen) in Germany
- GreenRE (Green Real Estate) in Malaysia
- BEAM plus (Building Environmental Assessment Method) in Hong Kong.

Initiators of these tools often claim that they have developed their own benchmarking tools because they felt that LEED and BREEAM did not suit their local built environment, climate, regulations and standards. As a result, they felt tailored tools were needed that also responded to some of the early-day critiques expressed about LEED and BREEAM (DGNB, 2009; HKGBC, 2013).

Yet, this claim is somewhat contradicted by the wide uptake of LEED and BREEAM around the world, which seems to imply that the tools are flexible enough for local adaptation. Further, it is striking that tools such as Green Star and the DGNB system are actively exported around the world by their developers. Green Star is also applied in South Africa (GBCA, 2012); while the DGNB system has been exported to some 20 countries such as Bulgaria, Thailand, China and Brazil (DGNB, 2013). Best-of-class benchmarking tools appear to have become a market in themselves, and to date limited attention has been paid to competition between such tools (exceptions from other fields are described in Cashore et al., 2004; Smith and Fischlein, 2010).

But not only are full buildings subject to classification and benchmarking. Building products and building users can be benchmarked and classified as well. Typical examples are the Green Labelling Scheme and ECO-Office in Singapore. The Green Labelling Scheme was launched in 1992 by the Singapore Environmental Council, an NGO, to endorse consumer products and services that have less undesirable effects on the natural environment. The Green Labelling Scheme can best be understood as an information-based tool that indicates whether or not a product meets the criteria set by the Singapore Environmental Council. In contrast to the earlier discussed benchmarking tools, the Green Labelling Scheme does not give a relative score. Products that have been labelled range from building materials, such as brick, ceramics and paints, to materials and products typically used in buildings, such as stationery paper, photocopiers and computers (Singapore Environment Council, 2013b).

The scheme is closely related to ECO-Office, which was launched in 2002 as a joint initiative between the Singapore Environmental Council and a major Singapore-based real estate developer. ECO-Office seeks to raise awareness among commercial property occupants (office users) about environmental and resource sustainability. One of the approaches for doing so is through an online office benchmarking tool. Participants in this tool can perform a self-audit based on metrics, such as corporate environmental policy and commitment, purchasing practice, waste minimization measures and the level of recycling. Credits are awarded to particular performance. A number of credits relate to the use of products that are considered as environmentally sustainable under the Green Labelling Scheme (Singapore Environment Council, 2013a). Property occupants that rate well enough, and undergo a third-party assessment, are awarded a label to indicate their performance. This label can, like the BREEAM and LEED ratings discussed above, be used for further promotional activities. The label differs from the benchmarking tools discussed above, however, as it indicates the performance of building occupants and not the building itself. This allows, in theory, occupants in poor performing buildings to nevertheless show positive behaviour in terms of environmental sustainability.

Another take on classification is the Sustainable Business Leader Program in Boston (SBLP, 2013). The programme was developed and implemented by a Boston-based NGO, the Sustainable Business Network of Massachusetts, which is financially supported by the city of Boston, in Massachusetts. The programme does not classify buildings, building products or building occupants, but those involved in designing and constructing buildings, among others. The programme was created to assist Boston-based businesses in becoming more sustainable. It offers these businesses technical, hands-on assistance that is affordable, practical and actionable.

The programme may best be understood as a classification tool combined with an educational programme. To be classified and recognized as 'sustainable business leader', participants have to fill out a questionnaire with a total of 118 questions about issues such as energy, water, transport and waste. The questionnaire is then reviewed by the administrating body, and based on this review the participants are audited. The audit seeks to gain insight into whether the participants have carried out their self-assessment correctly, and also to ask them about their ambitions in terms of environmental sustainability. As with the other classification and benchmarking tools discussed, participants receive a label indicating their performance, which can be used for marketing purposes. By the end of 2013, the programme had issued labels to over 130 local Boston-based businesses.

Governmental involvement in classification and benchmarking tools

It is of interest to pay some attention to the administrative bodies of benchmarking tools such as LEED and BREEAM, and their relationship with government. LEED is administrated by the US Green Building Council, a tax-exempt membership-based NGO. The council was established in 1993. Its constituency includes representatives from the building sector, government, NGOs and citizen representatives; while its board of directors includes representatives from the building sector and government (USGBC, 2013d).

The council is independent from US governments, but is indirectly supported by government. For instance, state and local governments throughout the United States offer developers and building owners financial incentives, such as tax breaks, for having their buildings LEED certified (USGBC, 2013a; see also examples discussed in Chapter 2). Also, governments throughout the United States have adopted LEED regulatory requirements in their policies, for instance, by requiring their buildings to meet certain LEED ratings. With 27 per cent of all LEED projects being government owned or occupied in the United States, governments are a major client for the council.

BREEAM is administrated by the Building Research Establishment. On conception, it operated as a departmental executive agency, before it was fully privatized in 1997. The BRE is owned by the BRE Trust, a charity dedicated specifically to research and education in the built environment. All profit made through BREEAM is used for educational and research purposes, overseen by this trust. In becoming independent from government in 1997, the BRE could assess and certify buildings, which was the final step needed to launch BREEAM (BRE, 2013c).

Although the BRE is now independent from government ties, it is still supported by the government. For instance, the UK Government requires all building on government estate to achieve the maximum BREEAM rating for new buildings and at least the second-best BREEAM rating for major refurbishments (BRE, 2013b). In addition, the UK Government hires the BRE for the development of construction codes, for instance, the code of sustainable homes (Communities and Local Government, 2006).

Aside from supporting the development and implementation of these tools by others, governments can, of course, also launch their own (voluntary) classification and benchmarking tools. Examples include:

- NABERS (National Australian Built Environment Rating System) in Australia (Office of Environment and Heritage, 2013)
- Green Mark in Singapore (BCA, 2012)
- Green Building Index in Malaysia (GBI, 2013)
- GRIHA (Green Rating for Integrated Habitat Assessment) in India (Ministry of New and Renewable Energy, 2012).

NABERS is an intriguing example. It is a classical benchmarking tool, which measures the environmental performance of buildings and tenancies in terms of energy efficiency, water usage, waste management and indoor environment quality. Buildings are rated on a scale from one to six stars, with six stars indicating market-leading performance. This rating reflects the performance of buildings in use, and the rating has to be renewed yearly.

NABERS was launched as a voluntary programme in 1998 by the state government of New South Wales. In 2000, NABERS was introduced throughout Australia as a national voluntary programme. As of 2010, under the Australian Building Energy Efficiency Disclosure Act 2010 (Australian Government, 2010), a NABERS rating is compulsory when commercial office space of 2000 square metres or more is offered for sale or lease. NABERS is further incorporated in CitySwitch Green Office, the partnership of governments and office tenants discussed in Chapter 3. Under CitySwitch tenants seek to achieve a particular NABERS rating for their tenancy. This combination of a voluntary programme that is mandatory in other governance tools may explain its current success. In 2013 more than 70 per cent of Australia's office space was NABERS rated (NABERS, 2013).

Green Mark in Singapore provides for a comparable story of a benchmarking tool developed and implemented by government. Green Mark was introduced as a voluntary programme in 2005 by the Singapore Building and Construction Authority. It seeks to drive Singapore's building sector towards developing more sustainable buildings. A Green Mark rating reflects a building's energy and water efficiency, environmental protection and indoor environmental quality. Buildings are rated Gold, Gold+ and Platinum to indicate comparative differences. To achieve a wide uptake of the tool a fund was started to financially support developers in achieving high Green Mark ratings.

As of 2008, a Green Mark rating of at least Gold has become a compulsory requirement for all new developments larger than 2000 square metres. As of 2013, a Green Mark rating of at least Gold has become compulsory for commercial buildings larger than 15 000 square metres, when they are retrofitted. With the financial incentives still in place and by increasingly mandating Green Mark rating to existing buildings, the Government of Singapore aims to have 80 per cent of its building stock Green Mark rated by 2030 (BCA, 2013; Green Mark, 2013).

4.2.2 Green Leasing

With classification and benchmarking tools moving increasingly towards rating buildings according to their actual performance when in operation, the role of building users becomes of significant importance. After all, if a tenant of a LEED Platinum certified building leaves the lights and heating on 24 hours a day the building will certainly not achieve its predicted energy savings.

At the same time, an office tenant may seek to reduce their energy consumption and improve their overall environmental performance as a part of their social corporate responsibility strategy, but may find there is only so much they can do without a retrofitting of the building it is leasing. The building's owner, however, may be unwilling to make the retrofits, because the financial advantages of a retrofit come to the tenant in the form of a reduced energy bill. This is an example of a split incentive, which was discussed in more depth in Chapter 1. For both examples given, green leases may provide a solution.

A green lease is, like any lease, an agreement between a property owner (landlord) and a tenant. In the agreement both parties' responsibilities are laid out. Normally, the landlord provides for a building space and will ensure that the space can perform a certain function, for instance, office space; while the tenant promises to pay rent for the space and use it in a certain way, for instance, only applying interior materials agreed upon. Leases are a familiar tool and the language for leases has crystallized and become standardized over a long period of time. Yet, it has done so for conventional buildings. For sustainable buildings the language for leases is not yet clear and has only recently started to develop (Brooks, 2008; California Sustainability Alliance, 2009; Kaplow, 2009).

Green leases seek to address the split incentive problem landlords and tenants face. In a green lease they can agree that the landlord carries out certain retrofits, but only if the tenant agrees with an increase in rent, or shares the 'profit' of the reduced energy costs with the landlord. They can agree that the tenant will only use specific interior designs that do not negatively impact the overall performance of the building, or that the tenant will use the building in an efficient and environmentally sustainable way. Green leases can help both the landlord and the tenant to come together and overcome existing split incentive problems. In working together they can save costs (Brooks, 2008).

Green leases have received much attention from governments and are currently being trialled in a range of voluntary programmes (for an overview, see Green Lease Library, 2013). A good example is the Green Leasing Toolkit in California, United States (California Sustainability Alliance, 2009). The toolkit is predominantly a website that brings together information about green leases. It explains the advantages of a green lease, helps organization to develop green leases, communicates policies on urban sustainability to the market and seeks to develop a language for green leases. Participants using the toolkit can compare their own performance against that of others.

Another example comes from Australia. Here a National Green Leasing Policy was introduced in 2010 (Government Property Group, 2010). The policy applies to new leases or lease renewals for buildings where government is the tenant, the leased area is 2000 square metres or greater and the lease term is two years or longer. The policy builds on the NABERS benchmarking tool discussed earlier. It stipulates a 4.5-star NABERS rating as requirements for the leases, out of a maximum NABERS rating of six stars. The policy seeks to develop a common language for green leases throughout Australia, and to stimulate the uptake of green leases in the private sector. Through the CitySwitch Green Office programme (see Chapter 3) non-governmental tenants are also stimulated to participate in the policy.

4.2.3 Private Regulation

Many of the benchmarking tools discussed above are examples of private regulation that seek to improve urban sustainability, and to a certain extent urban resilience. Other well-known private regulations are those developed by the International Organization for Standardization (ISO). The organization is perhaps better known in the building sector for the regulation and standards it has developed for building materials and construction processes, and more general standards such the ISO 14 000 environmental management system or the ISO 50 000 energy management system (ISO, 2012a).

Private regulation provides a common language for actors in the building sector. It may also inspire governments to develop comparable regulation for their own jurisdiction, or even mandate the use of originally voluntary private standards.

International Green Construction Code

The International Green Construction Code provides an intriguing example of a set of private standards, a code, which seeks to improve the environmental performance of buildings (International Code Council, 2010). The code was developed by a range of organizations based in the United States, among others, the International Code Council, the American Institute of Architects and the US Green Building Council. As with many of the examples discussed in this book, the code seeks to reduce the negative impact of the built environment on the natural environment. It is marketed as 'a regulatory framework for new and existing buildings, establishing minimum green requirements for buildings and complementing voluntary rating systems, which may extend beyond the baseline of the [code]' (International Code Council, 2013).

The code reflects the structure of statutory regulation, discussed in Chapter 2. It sets requirements for building designs, construction work and the use of buildings. It builds on both prescriptive and performance-based standards. For instance, the prescriptive section 402.2.1 on buildings in floodplains prescribes that 'building and building site improvements shall not be located within a floodplain' (International Code Council, 2010, p. 35); and the performance-based section 602.2.4 on energy use intensity requires that 'the building shall be designed and constructed to deliver an energy use intensity (EUI) that would place the building in the top 10 per cent of existing buildings in terms of energy performance' (International Code Council, 2010, p. 76). Both prescriptive and performance-based standards in the code build strongly on measurable metrics, which may ease the enforcement of the code.

The code can be incorporated in the mandatory construction codes of governments, or in the voluntary requirements set by classification and benchmarking tools. It appears to predominantly aim for adaptation by governments. Chapter 3 of the code addresses this exact issue. It provides a list of provisions governments can choose from for inclusion in their construction codes. It also provides a list of project electives that governments may consider necessary for a project to comply with the code and their local construction codes.

The code has been adopted by a number of jurisdictions in the United States (International Code Council, 2012). For instance, the state of Florida has adopted the code as an option for the retrofitting and new construction of all state-owned facilitates. North Carolina has adopted parts of the code, with amendments, in its mandatory construction codes. There appears to be no research available yet on the performance of the code.

Local regulation by citizens

The Aldinga Arts Eco Village, 40 minutes from Adelaide, Australia, shows that without government intervention, or even without the involvement of large and well-organized non-profits, citizens together can seek to develop and implement a set of regulations that addresses urban sustainability and resilience.

The Aldinga Arts Eco Village was established in 2003. It is a 33-hectare community development project based on progressive sustainability design principles. The village and its current 200 residents are subject to by-laws that were developed by its residents (Aldinga Arts Eco Village, 2003). The by-laws require the village's residents to be environmentally and socially responsible. Future residents are encouraged to communicate and consult with neighbours prior to purchasing land and prior to building their homes.

The by-laws state requirements to the orientation of buildings, their energy efficiency and rainwater harvesting. They stipulate the amount of open space in the village, the design of roads and their water runoffs, and prohibit solid fencing as a means to fence off private property. The by-laws stipulate that indigenous species are planted, weedy and invasive species are to be avoided, and encourage the planting of 'useful' plants – that is, those that provide food, fibre, shade or shelter. A central component of the village is a communal farm. The residents of the village expect that this farm will produce enough food by 2020 to make them self-sufficient.

It goes without saying that not many urban environments are characterized by this luxury of having 33 hectares of land to house and feed only 200 people, or one hectare for six people. A city with an average density, such as London, houses 60 people per hectare, while a city with a high density, such as Hong Kong, houses more than 300 people per hectare (Karathodorou et al., 2010). This does, however, not imply that citizens of London or Hong Kong cannot develop and implement their own regulations, seeking to improve urban sustainability and resilience. The international Transitions Town network provides guidelines and supports citizens in taking such action (Transition Network, 2013a).

The Transitions Town network aims to mobilize community action, and foster community engagement and empowerment around climate change. It seeks to speed up a transition towards a low carbon economy. Since its initiation in 2006, the network has expanded rapidly in and outside the United Kingdom (Seyfang, 2009). The network predominantly seeks to support individuals and citizens to

> be the catalyst in your community for an historic push to make where you live more resilient, healthier and bursting with strong local livelihoods, while also reducing its ecological footprint. (Transition Network, 2013a)

There is no predefined format of the initiatives that fit the network's ideology. Examples range from local food-growing groups where people grow food through garden sharing schemes, to transition street projects where communities work together to improve urban sustainability on a street-by-street basis. Through the network's website participants can download sample constitutions that may help them to draw up the regulations for their own initiatives (Transition Network, 2013b).

A 2009 study of 79 UK-based Transition Town initiatives predominantly reports on successes in terms of activating people to start up projects (Seyfang, 2009). Participants in these generally appreciate the communities built through the initiatives, and the links with other voluntary programmes and governments.

The study, however, reports limited actual outcomes of the Transition Town initiatives it studied. Participants report problems that relate to group dynamics and the difficulty to govern groups of citizens, and the lack of funding and time to make the initiatives a success. Finally, the analysis found that food and gardening activities were the most popular initiatives (such initiatives were started in 40 per cent of the Transition Towns it studied), and that energy-efficiency initiatives were significantly less popular (such initiatives were started in 10 per cent of the Transition Towns it studied).

Government support for and adoption of private standards

Governments are actively using private regulation in the governance of urban sustainability and resilience. The Government of Singapore, for example, financially supports small and medium enterprises to achieve ISO certification in areas such as environmental management. The Capacity and Development Grant funds up to 70 per cent of these enterprises' costs of obtaining such certification, relating to consultancy, manpower, training, certification, upgrading productivity and developing business capabilities for process improvement, product development and market access (SPRING, 2012). Other examples relate to the adoption of the International Green Construction Code, governmental support for the use of private benchmarking tools (see above) or even the adoption of the regulatory requirements underpinning LEED or BREEAM in local construction codes (Schindler, 2010).

Adopting standards developed by NGOs appears to be an easy and cost-effective way for governments to quickly introduce regulatory requirements that may help to improve urban sustainability and resilience. This strategy is not without risks (Corbett and Muthulingam, 2007; Schindler, 2010; Schmidt and Fischlein, 2010). Governments need to be careful in adopting private regulation as a baseline for their own construction codes, or even supporting the use of these. Private regulations emerge under a different set of accountability and legitimacy rules than public regulation. Although the administrative organizations behind, for instance, BREEAM, LEED and the International Green Construction Code represent a wide range of stakeholders, governments included, they do not have the democratic legitimacy governments normally possess.

4.2.4 Innovative Financing

Building owners or developers often find it hard to get mortgages for retrofitting or constructing buildings with high levels of environmental performance. Banks are still found to be risk averse and unwilling to finance sustainable buildings since they lack awareness of the financial benefits of such buildings (Managan et al., 2012; Pivo, 2010). Innovative tools have begun to emerge to address this problem. Interestingly, governments play a strong role in their development and implementation.

ESCOs

Energy Service Companies (ESCOs) are companies that provide energy services to their clients. The general business model is that the company installs energy-efficient measures in its client's buildings, operates and maintains these, and may even supply all non-generated energy its client needs. For doing so, the company charges a fee below the energy costs the client faced before modification. The business model assumes that the company will reduce actual energy consumption in such amounts that the fees it charges are profitable for the service it provides (for further details, see Vine, 2005). ESCOs are particularly popular in the United States, but have been introduced in a range of other countries as well (Goldman et al., 2005; Larsen et al, 2010).

In the Netherlands, for instance, the performance of ESCOs was experimented with in Rotterdam in 2010 (City of Rotterdam, 2011; Simons, 2013). The City of Rotterdam owns a number of swimming pools and considered ESCOs a positive approach to increase the energy efficiency of these, while saving costs at the same time. The city has entered into a ten-year contract with an ESCO for the energy management of its swimming pools. The ESCO has promised the city energy consumption reductions of 35 per cent as compared to the base year 2009. The ESCO does not charge the city for its service but will keep the financial savings it realizes of up to 35 per cent energy reductions, and will split the financial rewards of all additional savings with the city.

In its first year of operation the ESCO achieved 34 per cent energy reduction savings and it is expected that this number will grow because of the long-term ten-year contract.[1] This allows the ESCO to fully understand the operation of the swimming pools and improve their energy performance even more (Hofman, 2013).

Despite this positive outcome, less positive outcomes were also reported. The development of the contract was a relatively time-intensive process. The city, the organizations in the ESCO and Dutch legal firms had no prior experience with this new form of contracting energy service. This led to a long contract development process, which was expensive due to the relatively large number of legal and financial advisers involved. In addition, the City of Rotterdam has taken significant risks in the ESCO contract. It has agreed to ensure that energy costs will be paid to the utilities in case the ESCO is unable to do so. This risk is considered too significant by Dutch experts in the field, who therefore consider this form of contracting unlikely to be used for business-to-business contracting (DGBC, 2013).

The complicated process of developing this first ESCO contract in the Netherlands has, according to administrators involved in developing the contract, resulted in a situation where various parties in the Dutch building sector think this form of contracting is too complicated (Van der Heijden, forthcoming 2014b). The contract is, however, freely available to all parties interested in entering into ESCO contracts, which saves them the time and money involved in developing this first contract.

In Singapore, on the contrary, ESCOs are considered a successful approach in funding the transition towards more sustainable buildings. To increase the uptake of ESCOs throughout Singapore the government started an ESCO accreditation scheme in 2005. The scheme is administered by the Energy Sustainability Unit at the National University of Singapore, which receives funding from the national government (Briomedia Green, 2012). The scheme seeks to enhance the professionalism of ESCOs and the quality of services they deliver. By the end of 2013 the scheme had accredited close to 20 ESCOs in Singapore (E2PO, 2013).

In addition, the national government funds a large-scale programme, Solar Leasing, which supports households in installing solar panels on their homes through ESCOs. Solar Leasing applies to all of Singapore's publicly developed but privately owned 9000 housing blocks, housing the majority of Singapore's citizens. With this governmental funding it is anticipated that over the next two decades Singapore's housing sector will be a net producer of energy (Wong, 2011) at relatively low societal cost. The governmental funding is S$31 million and may see 4.4 million Singapore residents served. That is, for about S$7 per citizen (about US$5.50) Singapore households will no longer be dependent on imported energy.

Tripartite financing

Governments are in a unique position to support building owners and developers in generating funding for the development or retrofitting of sustainable and resilient buildings. After all, they are a low-risk client for mortgage suppliers and they have much experience in and well-developed administrative systems for collecting fees and payments. Governments make for a perfect intermediary in the relationship between financial institutions and building owners and developers. Recently, governments have indeed begun to take up this role as intermediary.

A typical example is the 1200 Buildings programme in Melbourne, Australia, which was launched in 2010. It seeks to speed up the retrofitting of commercial buildings in the city's central business district. The programme was developed by the city of Melbourne in collaboration with a national bank, a major fund manager and property owners. A key part of the programme is Environmental Upgrade Financing, an approach to financially supporting building owners that was not implemented anywhere in the world prior to 2010. It allows the city of Melbourne to levy a new form of statutory charge to which participants in the 1200 Buildings programme voluntarily agree.

Under the programme individual building owners commit to a minimum 38 per cent reduction of energy consumption in a letter to the Mayor of Melbourne. In return the city of Melbourne provides these building owners with funds necessary to retrofit their buildings to achieve this goal. The city of Melbourne lends the funds from a financier or bank. Through the new statutory charge introduced it collects funds from the building owner to repay the loan to the financier or bank. The programme further provides the building owners with information on how to retrofit their buildings, and it creates a community of building owners that can learn from each other's experiences. Participants may use a promotional logo to highlight their involvement in the programme (City of Melbourne, 2010).

By the end of 2013, 26 participants had joined the programme representing 45 buildings (City of Melbourne, 2013). The programme is internationally acclaimed (C40 Cities, 2013), but is not widely followed. Only the city of Sydney has introduced a comparable form of Environmental Upgrade Financing. Other cities seem to consider the complicated legal process of introducing a new form of statutory charge as a barrier (Van der Heijden, 2013b).

Another example is the Property Assessed Clean Energy (PACE) programme in the United States, introduced in 2008 (PACE Now, 2013b). Under the PACE programme local governments can issue bonds to investors and then use these funds as loans to homeowners and commercial property owners. The loans can only be used for energy retrofits of existing buildings and are repaid through additional property taxes charged on the building to which the loan relates. This attaches the loan to a building and not to an individual. With property taxes being lower than commercial interest rates for mortgages, building owners find a relatively inexpensive form of financing in PACE.

As with the Australian examples, PACE requires changes in state legislation to allow local governments to issue bonds. States in the United States appear willing to make such changes, and by the end of 2013 more than half of the states in the United States (31) had made these changes (PACE Now, 2013a). The performance of the programme is yet unclear (Zimring et al., 2013). The programme experienced a considerable blow from the subprime mortgage crisis in the United States. Responding to the crisis, the US mortgage authorities, Freddie Mac and Fannie Mae, refused to finance mortgages under the PACE programme. This in effect terminated the PACE programme for supporting homeowners (Bird and Hernandez, 2012; Sichtermann, 2011).

Revolving loan funds

Revolving loan funds are a source of money that is normally made available to support small and medium development projects. In particular, these funds seek to provide loans to individuals, organizations or projects that do not qualify for traditional loans. The loans normally do not fund full projects, but are a bridge between the loans a borrower can get on the market and the funds needed for a project. The funds are revolving because when loans are paid back to the central fund it can issue new loans to other projects (Boyd, 2013; CDFA, 2013; Indvik et al., 2013).

Revolving loan funds are particularly popular with universities and other educational institutions in the United States (Flynn, 2011; Foley, 2011; Indvik et al., 2013). Throughout the United States over 80 revolving loan funds were recorded in 2013, containing close to US$120 million (AASHE, 2013). The Billion Dollar Green Challenge is the largest of these. It is a voluntary programme that encourages educational institutions to invest a total of US$1 billion in a self-managed revolving fund to finance energy-efficiency upgrades of educational buildings (Green Billion, 2013).

The challenge was launched in 2011. By the end of 2013, 41 institutions had committed themselves to the challenge, and had invested close to US$80 million to the fund. By joining the challenge, participants do not only find financial support for their projects, they are also provided with information and best practices about how to increase the (environmental) sustainability of their existing buildings (for example, Sustainable Endowment Institute, 2012). The challenge has funded retrofitting projects at universities such as Harvard University, Stanford University and Boston University.

Another example of a revolving loan fund is the Amsterdam Investment Fund (*Amsterdams Investeringsfonds*) of the City of Amsterdam, the Netherlands. Half of the €140 million revolving loan fund (about US$200 million) is reserved for fund projects that aim to improve urban sustainability and resilience, and in particular projects that seek to reduce energy consumption or projects that increase the generation of renewable energy. The fund was originally installed in 2011 by the Amsterdam city council. To reduce the risk of the fund, 80 per cent of the loans are issued to low-risk projects, while the remaining 20 per cent of loans are issued to more riskier projects, which are innovative and push the boundaries of urban sustainability and resilience (City of Amsterdam, 2013).

The performance of the fund was evaluated in 2013. It was found that the fund does not perform as anticipated (Boonstra, 2013). Participation criteria were considered to be too strict. For instance, to be eligible for funding a development project's cost has to be at least €2.5 million, which in effect excludes many small-scale projects. The evaluation also noted that the goals of the fund are in conflict. On the one hand, it seeks to fund low-risk projects, while, on the other hand, it seeks to be a vehicle to aid the public interest in terms of urban sustainability and resilience. This may explain why after two years of implementation half of the funds available were not invested (Boonstra, 2013).

4.2.5 Contests, Challenges and Competitive Grants

Besides such innovations in project financing, governments and NGOs around the world are launching contests, challenges and competitions that seek to improve urban sustainability and resilience by funding projects and developing knowledge.

Contests and challenges

Contests and challenges build on the competitive nature of organizations and individuals. The Better Building Challenge in the United States, for example, supports commercial, industrial building and multi-family building owners by providing technical assistance and proven solutions to energy efficiency. Participants in the challenge commit to a 20 per cent energy reduction of their buildings or more over a period of ten years, and showcase the solutions they use. The challenge builds on transparent reporting aiming to develop best practices for others to follow. In doing so, the challenge seeks to reduce energy consumption in the United States and to stimulate the economy by raising the demand for building retrofits.

The challenge was launched by President Obama in 2011 and is administered by the US Department of Energy. Besides having support from President Obama, the challenge is interesting because it bridges building owners with financial institutions and utilities. Financial institutions have made a public commitment of close to US$2 billion in private sector capital to support energy upgrades of participants' buildings. Utilities have committed to support their clients in managing their energy consumption (US Department of Energy, 2013a).

By mid-2013, 7700 buildings were participating in the challenge, representing 2 billion square feet of building space. On average they had reduced their energy consumption by 2.5 per cent, representing a total savings of US$58 million. A considerable number of buildings, 1300, had already achieved the 20 per cent energy reduction goal, while another 2100 buildings had achieved energy reduction goals of 10 per cent (US Department of Energy, 2013b). The US Department of Energy rewards leaders in the programme by highlighting their best practices; for instance, in 2013 it issued about 20 press releases to highlight the successes of the challenge and its participants (US Department of Energy, 2013a).

Challenges are also organized on a local level. The Chicago Green Office Challenge is a business-to-business competition that encourages businesses in the city of Chicago, Illinois, to complete challenges that improve their environmental performance, to educate their employees and to share information about best practices. The programme was launched in 2009 by the city of Chicago in collaboration with ICLEI (the Local Governments for Sustainability network, discussed in Chapter 3), and seeks to make the transition towards higher levels of environmental sustainability 'fun' (ICLEI, 2009).

Examples of challenges are within-office or office-to-office competitions on energy consumption reductions. Participants in these challenges report their performance on a website and in doing so keep track of how they rank against other participants. A yearly awards ceremony highlights the leaders in the challenge and through press releases the city of Chicago seeks to highlight the successes of the challenge and its leaders further. In its first year participants in the challenges on average reduced their energy consumption by 8 per cent and their water consumption by 5 per cent.

A different approach to challenging businesses and individuals to show leadership in terms of urban sustainability and resilience is taken by the city of Boston, in Massachusetts, in the E+Green Building programme. The programme builds on a tradition in the building sector to award commissions for building design and development to the winner of a design contest (Nasar, 1999).

E+Green Building challenged designers and developers to propose a building design for one of three normally highly sought after sites in the inner-Boston neighbourhoods. In the E+Green Building programme the sites were assigned to the winning proposal. The design contest criteria were that proposals have to meet a maximum LEED rating of 'Platinum' (LEED is discussed above) and have to be energy positive ('E+'). The latter criterion implies that buildings should generate more energy than they consume (City of Boston, 2013b).

Through the programme the city of Boston hopes to demonstrate that it is possible to develop houses that are energy positive, and that it is financially viable to build such houses. The programme was launched in 2011 and attracted 14 design teams. Three of these will be built on the sites provided by the city of Boston. The city of Boston uses the various submitted designs as well as the experiences with the three awarded designs as demonstration projects in exhibitions and lectures (City of Boston, 2013a).

Competitive grant funding

Where in the above examples participants of contests and challenges are rewarded with, predominantly, media exposure, development projects, information on how to develop or retrofit buildings with high levels of (environmental) sustainability, and sometimes access to funds, another reward for participating can be actual funds for developing or retrofitting buildings. This is in a nutshell what competitive grant funding seeks to achieve (Irvine et al., 2012).

Competitive grant funding builds on the assumption that if participants are stimulated to compete with each other for funds they will seek to achieve more ambitious outcomes than when given a lump-sum subsidy for a prescribed outcome. Competitive grant funding stands next to traditional subsidies as prescriptive regulation stands next to performance-based or goal-oriented regulation (these forms of regulation and subsidies are discussed in Chapter 2).

The Grassroots Program in the Boston, is an example. The programme funds the design and construction of community gardens and open spaces; and it conveys city-owned land to non-profit organizations for community benefit and use. The programme was launched in 1985 and has since awarded over US$20 million to more than 170 community gardens, urban agriculture projects and open space initiatives (City of Boston, 2010).

The Building Innovation Fund, in the state of South Australia, is a comparable initiative. This A$2 million fund was established by the Government of South Australia in 2008 to demonstrate innovative ways to reduce the carbon footprint of existing commercial buildings. The fund was closed in 2012 after four yearly rounds of funding. The fund has offered grants to owners of office buildings, hotels and shopping centres for initiatives that demonstrated new and leading-edge approaches to reducing a building's energy use and greenhouse gas emissions. A total of 11 projects were awarded a grant (Government of SA, 2012).

The awarded projects predominantly seek to test the feasibility of new approaches to retrofitting existing buildings, for instance, the installation of a tri-generation plant in an existing building or the installation of living walls and green roofs. Administrators of the grant are particularly positive to this type of funding. They consider that it has spurred innovation among participants in the funding rounds to develop proposals that not only aim for 'business-as-usual' retrofitting but also seek to push the boundaries of technological solutions that are financially viable.

The Government of South Australia has received many more proposals that highlight innovative solutions to retrofitting existing commercial buildings than the number of grants awarded. It now has a database of (proposed) solutions and a (smaller) database of best practices realized with grant funding. A number of these are accessible through a website that the Government of South Australia maintains. Also, some of the projects that have not received funding are still realized by their private sector initiators. The fund may be considered a means to make actors in the building sector think beyond business-as-usual practice, and make them realize and utilize the potential of current technologies they would otherwise not have done (van der Heijden, 2013d).

4.2.6 Intensive Behavioural Interventions

All of the above voluntary programmes and market-driven governance tools have a dominant focus on commercial buildings. Residential buildings account for roughly half of the greenhouse gases produced in the building sector in developed economies (Greenhouse Office, 1999). Seeking voluntary improvement of residents' environmental performance, governments have developed 'intensive behavioural interventions' (a term that I borrow from the field of psychology; see Howlin et al., 2009).

These interventions are comparable with more traditional information campaigns by governments (Stewart, 2006). They differ from these, however, in that the interventions seek direct engagement with households and individuals and are highly active forms of information supply. Traditional information campaigns are often passive forms of information supply, for instance, a newspaper advertisement campaign that informs households on how they can reduce their energy consumption.

In particular, Australian governments have been very active in informing households on how they can reduce their energy and water consumption, and in doing so save costs (KPMG, 2008). The ClimateSmart Home Service in the state of Queensland that ran from 2009 to 2012 is a typical example. It built on household audits by a licensed electrician. The electrician would carry out a household audit based on their energy bills, and provide advice on how the household could reduce its energy consumption. In addition, the electrician would install a number of energy-efficient light globes, a water- and energy-efficient showerhead and a wireless energy monitor to help track electricity usage around the home. Households were charged a fee of A$50 for the service, while the value of the audit and products was about A$450. The programme has delivered 340 000 services throughout the state of Queensland, which is but a small percentage of the state's 1.7 million households (about 20 per cent). In 2012 the programme was terminated by the Liberal-Conservative party, who took over political power from the Liberal party in March 2012 (Hurst, 2012). It considered the programme too much of a burden for taxpayers (see also Chapter 3).

A comparable Australia-wide programme is the Home Energy Saver Scheme (Australian Government, 2013a). This programme seeks to support particular groups, such as low-income households and pensioners.

A somewhat different approach to making individuals aware of and improving the (environmental) sustainability of buildings is through education. The Singapore Energy Experience Programme seeks to educate secondary school students (Singapore Power, 2013). The programme aims to make students aware of the importance of energy conservation and the significant roles that electricity plays in their everyday lives, that of their families and in society more broadly.[2] As a part of the programme students are asked to carry out a home electricity audit.

Educating (young) students about energy and resource efficiency and the role of the built environment appears a popular approach around the globe (Dawe et al., 2005). It is expected that if people are educated on such matters at a young age they will be more aware of energy and water consumption and waste production, and behave in more desired ways. It is further expected that children may be the catalysts of change when they inform their parents about the importance of a more sustainable lifestyle.

However, studies on the effects of such educational programmes point in different directions. While some trace increased awareness and increased behaviour of students and their parents (Armstrong, 2004; Taber and Taylor, 2009; Zografakis et al., 2008), others do not or show that initial effects fade out over time (Henryson et al., 2000; Jurin and Fox-Parrish, 2008).

4.2.7 Sustainable Procurement

Besides participating in a voluntary programme or market-driven governance tool, organizations and individuals can, of course, use their purchasing power as consumers to seek an increase in urban sustainability and resilience. Around the globe governments and businesses have pledged to do so and have introduced purchasing procedures that stipulate that, for instance, their offices need to meet particular sustainability criteria.

Such purchasing procedures have become known as sustainable procurement (Australian Government, 2013b; Forum for the Future, 2007; Preuss, 2009; UNEP, 2008b; Walker and Brammer, 2009; Walker and Phillips, 2009). Sustainable procurement is, for instance, a recurring topic on the policy agenda in India (TERI, 2008). In 2012 the Public Procurement Bill 2012 was passed. The bill stipulates that environmental criteria of a product or service may be adopted as one of the criteria for the evaluation of a tender (Government of India, 2012). The bill may be considered one step closer to the introduction of sustainable public procurement criteria that started with the introduction of the Energy Conservation Act in 2001. This act sets criteria for the energy efficiency of products and services purchased by Indian governments (Perera et al., 2007).

In the Netherlands, all Dutch governments together spend about €60 billion, or 10 per cent of the Dutch gross domestic product (GDP), on goods, (building) projects and services. In applying sustainable public procurement criteria (*Duurzaam Inkopen*) the Dutch governments together can influence the market for sustainable products and development, and so affect urban sustainability (Dutch Government, 2013).

The objective of sustainable public procurement is to achieve 100 per cent sustainable procurement by Dutch governments by 2015. The Dutch Ministry of Infrastructure and the Environment, together with other public authorities, has developed a series of purchasing criteria that Dutch governments can apply in their tendering and procurement processes, and can include in contracts with suppliers.

One of the critical aspects of developing these criteria was to ensure that they conform with European tendering processes. After all, sustainable procurement criteria may require local products to reduce carbon emissions from transport, while the aim of the European Union is, among others, to allow for a free flow of products and services and a reduction of trade barriers. Sustainable procurement criteria may then clash with the ideology and European tendering procedures because they introduce new trade barriers. The European Commission has introduced a few sustainable procurement criteria in European legislation, but these all relate to social aspects of sustainability and not to environmental aspects (European Commission, 2013).

An indirect goal of sustainable (public) procurement may be to support the introduction of particular products, services or even the government tools introduced above. Governments in Australia, for example, are now actively requiring specific Green Star ratings when they seek to renew a lease or commission a new development project (Van der Heijden, 2013b). In doing so they have been acting as a 'launching customer' for this benchmarking tool (Hofman and De Bruijn, 2010; Van der Horst and Vergragt, 2006).

Another example is New York City's government that seeks to accelerate a transition towards urban sustainability and resilience. By transforming its own buildings the government aims to change the market through demand, as well as by providing best practices and lessons on retrofitting existing buildings in the city (Kingsley, 2008). By passing Local Law 86 in 2005 the city requires that building projects receiving more than a specified amount of city government funding achieve a LEED rating level of Certified or Silver (a mid-tier rating), or reductions in energy cost and potable water use (City of New York, 2005). Since 2011, the New York City Department of Housing Preservation and Development requires that housing construction and substantial rehabilitation of housing projects that receive funding from this department align with LEED standards.

4.3 CONSIDERATIONS

The real-world examples of voluntary programmes and market-driven governance tools addressed confirm the literature on these approaches to governance discussed in the first section of this chapter. Even more than collaborative governance, these approaches can be considered highly pragmatic and targeted to local circumstances in how they address urban sustainability and resilience.

Some of the tools seek to address problems on a small scale such as the Aldinga Arts Eco Village in Australia or the E+Green Building competition in Boston, MA. Others seek to address problems on a large scale such as the best-of-class building benchmarking tools BREEAM and LEED that are applied throughout the world, or the Transition Towns network that seeks to stimulate citizens around the globe to take local action in improving urban sustainability.

The focus of these tools appears less on learning and sharing lessons as was the case with collaborative governance (discussed in Chapter 3). Voluntary programmes and market-driven governance tools appear more interested in providing its participants with clear and predictable results – that is, reduced energy consumption and related reduced costs, or high marketability of their buildings as a result of a high ranking in a best-of-class benchmarking tool.

Overall, it may be concluded that voluntary programmes and market-driven governance tools are highly popular in addressing, predominantly, urban sustainability, and in some cases urban resilience. But, as with direct regulatory interventions and collaborative governance, these approaches to governance come with some drawbacks.

4.3.1 Some Concerns About Voluntary Programmes and Market-driven Governance for Urban Sustainability and Resilience

The broader literature that assesses the outcomes of voluntary programmes and market-based governance is all but positive. It repeatedly reports that the outcomes of this approach to governance do not surpass those of other approaches to governance, and repeatedly it reports that participants in, especially, voluntary programmes do not outperform non-participants (see further below). This may hold partly true for the examples studied in this chapter as well. For example, the tool that has performed best in absolute terms, LEED, only shows a marginal impact on urban environments when these numbers are looked at in relative terms (see a comparable argument in Hoffman and Henn, 2009).

By the end of 2013 the US Green Building Council announced that in the United States more than 10 billion square feet of built space was LEED certified (USGBC, 2013d). In India the number is astonishing as well: 1.8 billion square feet of built space was already LEED-India certified by the end of 2013, and the Indian Green Building Council expects that the country will soon overtake the United States in terms of having the highest volume of LEED certified built space of any country in the world (IGBC, 2013).

These are of course laudable results. But what do they really mean?

The current built-up space in the United States was by the end of 2013 assumed to be close to 350 billion square feet.[3] This dwarfs the success of LEED in seeing less than 3 per cent of all buildings in the United States certified over a 20-year period, or an average of about 0.15 per cent per year. The numbers for India are somewhat better, but still modest. Its current built-up space is, conservatively, estimated as 25 billion square feet,[4] giving LEED-India a coverage of under 6 per cent over the course of seven years, an average of about 1.15 per cent per year.

Urban environments in an economy such as the United States grow by about 2 per cent per year. This implies that since LEED was introduced in the United States its building stock has grown approximately 30 per cent, or 80 billion square feet. In other words, for each square foot of LEED certified space about eight square feet of 'not so leading' space was constructed. Interestingly, it is exactly the US Green Building Council that, rightfully, keeps stressing that high-performing sustainable buildings do not have to cost more than conventional buildings (USGBC, 2010). One would therefore expect that it seeks it members to achieve the highest environmental performance possible. Yet, whether or not the US Green Building Council does encourage that, only 6 per cent of LEED certified buildings is rated Platinum (the highest rating, indicating top performance in terms of urban sustainability). This shrivels the 'success' of LEED in terms achieving high-performing sustainable buildings to a mere 0.18 per cent of all buildings in the United States over a period of 20 years.[5]

What this example shows is that even under very low urban development and transformation rates, such as in developed economies, the impact of leading voluntary programmes and market-driven governance tools is limited. This may hold true all the more in rapidly developing economies where these growth rates are significantly larger.

It is, of course, questionable whether such tools should be assessed as successful or not based on 'hard outcomes' such as buildings built, or 'soft outcomes' such as lessons learned (Lyon and Maxwell, 2007; Reid and Toffel, 2009; Rogers and Weber, 2010). These tools may very well have important diffusion effects: the information on these and the experiences with them will gradually diffuse among participants and non-participants. As a result, non-participants may learn from these tools without actually joining them. This may result in improved performance of non-participants and, related, a diminishing of the differences in performance between participants and non-participants over time.

Taking LEED as an example again, between 2002 and 2013 it was mentioned in close to 250 articles in the *New York Times*, whereas in that same period the newspaper discussed building retrofits in close to 550 articles (corrected for articles that discussed building retrofits through LEED).[6] With building retrofits taking up about 98 per cent of the problem of urban unsustainability (IEA, 2009), it may be argued that this is extraordinary media coverage for LEED, especially given the tool's limited market coverage. This media coverage will no doubt greatly help LEED to become known far beyond its current participant base.

Other examples of soft but relevant outcomes of voluntary programmes and market-driven governance tools are: the testing of new methods of production, for instance, the capturing and use of grey water as in the example of the Greywater Guerrillas (discussed in Chapter 1); the setting of a focal point by governments or leaders in the industry of what can be achieved in, for instance, urban sustainability, such as the E+Green Building competition; or the establishment of peer networks, for instance, the various groups of professionals on the social networking website LinkedIn that have developed around many of the examples discussed in this chapter.

4.3.2 Potential to Overcome the Three Main Governance Problems

In Chapter 1 I introduced three of the most significant governance-problems that may stand in the way of increased urban sustainability and resilience as:

- Governments are slow to react to existing urban problems. It often takes a long time to develop and implement legislation and regulation, and even longer for these to cause their effects.
- Introducing new legislation and regulation is often inconsequential. In developed economies cities develop too slowly for new legislation and regulation to be meaningful. In developing economies cities develop too rapidly for new legislation and regulation to be meaningful.

- A number of market barriers stand in the way of capitalizing on the economic benefits that more sustainable and resilient cities can bring.

Whether and how are voluntary programmes and market-driven governance able to overcome the main governance-problems discussed in Chapter 1?

At first glance they appear to hold much promise for all of the three governance problems. In terms of addressing the existing building stock, the tools addressed have made considerable progress. The Amsterdam Investment Fund, the South Australian Building Innovation Fund, 1200 Buildings in Melbourne, Environmental Upgrade Agreements in Sydney and the PACE programme in the United States are but a few of the examples discussed that seek to stimulate property owners to voluntarily improve their buildings beyond governmental requirements. These appear to be tools that are relatively easy to duplicate in urban environments around the globe.

The tools studied appear to have a lesser focus on the residential building stock. It may be that commercial property has an advantage of scale over residential property, when considered from a property owner's perspective. A, say, 10 per cent cut of the electricity bill adds up to a large number for a large office tenant in the United States, while it may be less than US$10 per month for a household (EIA, 2013) or a small or medium-sized business, for that matter. Such a small saving will likely be considered as futile in the household's finances (Cialdini, 2009), which may take away the attractiveness of the financial rewards that market-based governance tools build on.

In addition, a number of the voluntary programmes discussed have a focus on rewarding and showcasing leadership, such as the Chicago Green Office Challenge and the PACE programme. Being rewarded for and having their leadership showcased may, again, be less attractive to households and small and medium-sized businesses alike, although the Boston Sustainable Business Leader Program is an example that has been successful in addressing small and medium-sized businesses.

The tools studied further appear to have a lesser focus on improving the resilience of the (existing) building stock. It may be more difficult to stress the financial rewards for voluntary retrofits to improve the resilience of buildings. The investments in retrofits for urban sustainability are clearly represented on the utilities bill, but not so the investments in retrofits for increased resilience. These investments will only 'pay themselves back' when and if disaster strikes. But, the chances that it does are unclear and are likely projected into the distant and unknown future by building owners (Cialdini, 2009).

It is therefore questionable whether voluntary programmes and market-driven governance tools are able to change behaviour without the presence of an imminent risk.

An illustrative example is the waning and waxing attitude of Australians to conserve water. Throughout the years 2002 to 2007 the country experienced extreme droughts. In the same period Australian households, generally, became more water efficient, particularly by a changing use of water in their gardens (Australian Bureau of Statistics, 2010). After 2008 Australia experienced fewer droughts (and even severe rainfalls). It appears that without the imminent risk of water shortages in play Australian households have also become less conservative in their water usage (Australian Bureau of Statistics, 2013). Nevertheless, the long-term risk of water shortages remain in Australia (Corck, 2010).

A number of the tools studied seek to incentivize developers of new buildings to achieve higher levels of urban sustainability than that required by governmental building regulation. In particular, the wide series of best-of-class benchmarking tools appears highly successful. Around the globe a wide range of such tools is now in place. In particular, these tools address commercial property (such as offices, hospitals and schools), and lesser so, residential property.

Again, the advantage of scale for commercial property, and the possibility to showcase leadership by commercial property owners and tenants may be explanatory here. In particular, the possibility to showcase 'green credentials' as part of a corporate social responsibility policy may be leading in commercial property owners' and tenants' demands for office space with high levels of urban sustainability (Arora and Gangopadhyay, 1995; Baron and Diermeier, 2007; Feddersen and Gilligan, 2001). In addition, absent of ambitious sustainable building regulation in, particularly, (rapidly) developing economies (Hong and Laurenzi, 2007; Managan et al., 2012), these types of governance tools may likely fill these gaps (Blackman et al., 2013).

This type of governance tool appears promising in the transition towards meaningful urban sustainability, but there are some caveats. In absolute terms the numbers of buildings built and retrofitted that best-of-class benchmarking tools have achieved are astonishing, but in relative terms they appear less promising in generating a significant change in urban sustainability.

It is also questionable whether the development of voluntary programmes and market-driven governance tools is subject to similar accountability and legitimacy requirements as the development of governmental regulation (Schindler, 2010).

Lastly, voluntary programmes and market-driven governance tools are generally critiqued for having weak enforcement regimes in place (Darnall and Carmin, 2005; Delmas and Keller, 2005; Rivera and de Leon, 2004). I am not aware of any study that has yet studied enforcement practices in tools such as LEED or BREEAM, but based on studies that address the enforcement of building regulation more generally, it may be expected that enforcement is the soft spot of these tools (Imrie, 2007; Van der Heijden, 2009). This means that these tools should be treated with care, particularly in contexts with a history of weak enforcement of governmental regulations.

The tools addressed appear highly suitable for addressing the wicked set of market barriers discussed Chapter 1. Individually, the tools appear to focus on distinct barriers. For example, BREEAM and LEED address first-mover disadvantages; the California Green Leasing Toolkit and the Australian National Green Leasing Policy address the split incentives between property owners and tenants; the various tools around ESCOs address the behaviour of building occupants; and 1200 Buildings in Melbourne and the Amsterdam Investment Fund address the 'vicious circle of blame'. Especially the links between these tools and the collaborations discussed in Chapter 3 seem promising. For instance, a particular level of LEED certification is a requirement in a number of the tools in the United States. Introducing such a measurable requirement may overcome the potential danger of the tools being nothing more than greenwash (Lyon and Maxwell, 2006).

Another promising insight from these examples is that some (city) governments are actively involved in linking and assembling these tools. The city of Sydney, for example, actively uses green leases, requires Green Star certification for its office space, supports strata owners through Green Strata and office tenants through CitySwitch (both discussed in Chapter 3), and collaborates closely with and supports commercial property owners through Environmental Upgrade Agreements and the Better Building Partnership (discussed in Chapter 3). By linking tools, and their administrators and participants synergies may be created that make the whole of these tools larger than the sum of its parts.

4.3.3 Promises for Governing Urban Sustainability and Resilience

In conclusion, the wide set of voluntary programmes and market-based governance tools appears promising for achieving, in particular, urban sustainability, but less so for achieving improved urban resilience.

The tools allow for highly targeted interventions that can address small and large governance problems on various scales, from highly local to international levels. The tools may easily exist side by side (Smith and Fischlein, 2010), and may even be able to reinforce each other. The drawbacks of this approach to governing urban sustainability and resilience relates to potential issues with enforcement and a lack of meaningful outcomes.

In developing these tools a choice will have to be made between relatively strict or more lenient participation criteria. It is likely that too strict participation criteria disincentivize prospective participants from participation. However, without some strictness in participation criteria it is unlikely that a tool will result in meaningful outcomes – for instance, the number of building built with high levels of environmental performance (Potoski and Prakash, 2009). That is, if a simple 'cheat sheet' explains how easily it is to achieve a relatively high level of certification under LEED it is questionable what this level of certification actually means (Seville, 2011).

5. Trends in and design principles for governance for urban sustainability and resilience

The previous chapters have highlighted that there is no shortage of governance tools that governmental and non-governmental actors alike have implemented in seeking to achieve urban sustainability and resilience. In mapping and describing close to 70 different tools from Australia, Germany, India, Malaysia, the Netherlands, Singapore, the United Kingdom and the United States it has, however, become clear that the performance of these tools is highly context specific.

A tool that shows considerable successes in one context, for instance, ESCOs (see Chapter 4) in Singapore, was found to fail in another context, for instance, in the Netherlands. Also, whether a tool achieves desirable outcomes depends on the problem it seeks to address. For instance, best-of-class benchmarking tools (see Chapter 4) were found to have achieved considerable successes in the commercial sector, but not so in the residential sector. Thus, it would be a futile undertaking to try to design a single optimal governance tool that is applicable in a wide range of contexts and is able to address a wide variety of problems (this is a recurring conclusion in the literature, see, for example, Gunningham and Grabosky, 1998; Wurzel et al., 2013).

The previous chapters have, however, provided me with a sufficient base to distil a series of design principles for governance for urban sustainability and resilience that may guide governments, businesses and civil society groups and individuals to develop governance tools targeted to their specific contexts and problems. Although these will not guarantee desired outcomes, they will help those involved in designing governance for urban sustainability and resilience to ask the relevant questions and assess their designs for tools against a set of principles that have a strong theoretical and empirical base.[1] In the second half of this chapter, I discuss what I consider to be the 12 most important design principles for the successful governing of urban sustainability and resilience. These principles will most likely be of interest to policy makers and practitioners.

Before I move to discussing these design principles I first wish to distil the major trends in contemporary governance for urban sustainability and resilience from the large number of governance tools discussed in the previous chapters. While this sample is not perfectly representative for all of the world's countries and the governance tools in these, I feel it is sufficient to display relevant patterns that may help to better understand the current and possibly future direction of governance for urban sustainability and resilience. I seek to uncover the dominant actors in designing and implementing the various traditional tools (Chapter 2) and innovative tools (Chapters 3 and 4) discussed, the scope of these tools, their goals and their approach to achieving these goals.[2]

5.1 TRENDS IN GOVERNANCE FOR URBAN SUSTAINABILITY AND RESILIENCE

When comparing the close to 70 governance tools discussed in Chapters 2 to 4, some trends come to the fore. These are:

1. There is much innovative governance activity, but the lion's share of governance activity is still traditional.
2. Governments have far-reaching roles in innovative governance tools.
3. There is much innovative governance for new commercial property, and less for existing (and) residential property.
4. There is much innovative governance for urban sustainability, and less for urban resilience.
5. Much innovative governance seeks to stimulate the uptake of innovative technologies, it lesser so seeks to stimulate behavioural change.

5.1.1 Trend 1: Much Innovative Governance Activity, but the Lion's Share is Still Traditional

Throughout the book I have discussed 16 specific examples of direct regulatory interventions, 18 examples of collaborative governance and 34 examples of voluntary programmes and market-driven governance.[3]

These numbers do not, however, imply that traditional approaches to governing urban sustainability and governance have given way to more innovative approaches. By and large, the built environment, and in particular urban sustainability and resilience, is still governed through direct regulatory interventions, with statutory regulation being the most common form of governance in this area (Baer, 1997; CHBA, 2001; Hansen, 1985; Imrie, 2004; Listokin and Hattis, 2005; VCEC, 2005). One only has to look at the construction codes cited in Chapter 2 to get a grasp of the reach of these traditional tools as compared to the rather limited reach of more innovative governance tools discussed in Chapters 3 and 4.

That is not to say that all direct regulatory interventions in this area are still traditional; nor that all collaborations, voluntary programmes and market-driven governance tools are highly innovative.

As became clear in Chapter 2, governments around the globe are now seeking to improve traditional direct regulatory interventions such as statutory regulation, subsidies, taxes and other economic instruments. They partly do so by combining different incentives in individual tools.

In Europe governments are combining mandatory requirements for energy performance of residential property with information on how this can be achieved in energy performance certificates. Such information may help consumers to better understand how they can actually take action in an area that is, often, far beyond their expertise. Interestingly, the lack of information about cost savings through higher levels of sustainability on the European energy performance certificates is considered one of their weak aspects (Gram-Hanssen et al., 2007).

In the United States and Australia governments are introducing smart taxes and building incentive programmes that give property tax reductions to highly sustainable buildings, or fast-track building proposals for these buildings through the permitting process. These incentives financially reward (future) property owners of highly sustainable buildings.

Singapore and India have introduced innovative approaches to sustainability and energy performance requirements. Singapore is an inspirational example of a holistic approach to regulating sustainable construction. Where most countries still have a one-aspect-at-the-time approach to sustainable construction in their statutory codes, Singapore seeks to stimulate their property owners to think holistically about issues such as a building's energy performance, its indoor climate and its water consumption. India's experimental regulatory approach to building energy performance, finally, is inspirational because it considers that different aspects of building energy performance (building envelope, large installations, small installations) have different lifespans and ask for different specialisms, which relate to different 'bundles' of regulatory requirements and thus different regulatory assessment processes.

The collaborations, voluntary programmes and market-driven governance tools discussed in Chapters 3 and 4 showed much overlap in structure with statutory regulation. They often build on a set of rules that participants are expected to comply with, compliance is to a lesser or greater extent monitored, and non-compliance is disciplined or compliance is rewarded.

Thinking of governance as a structure of regulating, monitoring and disciplining or rewarding behaviour is very deeply embedded in our way of seeking to achieve particular goals, which echoes the socio-legal perspective on governance discussed in Chapter 2, but this structure comes with a deeply embedded shortcoming: enforcement.

The enforcement of voluntary programmes and market-driven governance tools is considered a major weakness of these approaches to governance. The current book has not carried out any new research into the levels of compliance with the examples discussed, and very limited research is available on the enforcement practices and levels of compliance within these examples. The literature from other fields cited, however, raises some considerable questions about the effectiveness of voluntary programmes and market-driven governance tools in the absence of strong enforcement. The early-day critique of tools such as LEED and BREEAM discussed in Chapter 4 emphasizes this point.

Enforcement and, related, compliance was also found to be a major weakness to direct regulatory interventions in Chapter 2. Construction incidents around the world give insight into the difficulty of guaranteeing structurally safe buildings, both in developed and rapidly developing economies.

Various causes appear at play, but the high financial compliance costs of buildings and the high financial gains of non-compliance may very well be a major cause of non-compliance in the sector (Circo, 2008; Evans et al., 2009; Sohail and Cavill, 2008). Buildings are becoming more (technological) complex, aiming to achieve high levels of environmental sustainability. As a result, compliance with regulations related to, particularly, sustainability, will become more complicated to monitor. To take an extreme event, a building collapse as a result of non-compliance with structural safety regulation is highly visible, but non-compliance with energy performance requirements is less so. It is therefore of importance that strategies for monitoring compliance are included in the design and implementation of both traditional and innovative governance approaches (see 'design principle 2', below, and Chapter 6).

5.1.2 Trend 2: Governments Have Far-reaching Roles in Innovative Governance Tools

It has become clear that governments rely strongly on businesses and civil society groups and individuals in the development and implementation of direct regulatory interventions. This blurring of the public and the private came to the fore in Chapters 3 and 4 (see also Gunningham, 2009). Governments are actively involved in innovative approaches to governance such as collaborations, voluntary programmes and market-driven governance tools. Even more, by unpacking the roles of governmental actors in the 52 examples of collaborations, voluntary programmes and market-driven governance tools it becomes clear that governments have far-reaching roles in these. In Chapter 3 these roles were identified.

They are the:

- Initiating or leading of the development and implementation of innovative tools.
- Assembling of these to achieve synergies between innovative tools or between these and direct regulatory interventions.
- Enforcing or guarding of the implementation of these innovative tools.
- Supporting or facilitation of these innovative tools.

Table 5.1 provides an overview of which role governments have taken up in each innovative governance tool discussed.

Table 5.1 indicates that in 37 out of the 52 innovative governance tools discussed in Chapters 3 and 4 governments have initiated their development or led their development and implementation process (71 per cent). Although many of the best-of-class-benchmarking tools, for example, are now fully administrated by NGOs or businesses, often their development was initiated by a governmental organization. BREEAM (see Chapter 4) is a typical example.

In 39 out of these 52 innovative tools governments have sought synergies between the innovative tools by connecting them (75 per cent). For instance, the city of Sydney has bridged tools that target property owners with those that target building occupants. In other examples governments have sought synergies between the innovative tools and existing direct regulatory interventions. An example is the inclusion of meeting a particular rating in best-of-class benchmarking tools as sufficient for complying with statutory sustainable building regulation, as is done in various local jurisdictions in the United States. Another example is the fast-tracking of buildings that meet a particular rating in voluntary best-of-class benchmarking tools in the statutory construction permit process (both examples from Chapter 2). By seeking such synergies governments may aim to achieve a situation where the total of all governance tools (both traditional and innovative) is larger than the sum of them.

Table 5.1 Government involvement in the innovative governance tools discussed

Type	No.*	Role of government in governance tools**			
		Initiating/ leading	Assembling	Enforcing/ guarding	Supporting/ facilitating
Collaborative governance	18	12	13	12	16
• Government-led collaborations	10	10	10	10	10
• Private sector-led collaborations	4	1	1	1	3
• Civil society-led collaborations	4	1	2	1	3
Voluntary programmes and market-driven governance	34	25	26	22	32
• Classification and best-of-class benchmarking	11	8	10	4	11
• Green leasing	2	1	2	1	2
• Private regulation	4	1	1	1	2
• Innovative financing	8	7	7	7	8
• Contests, challenges, competitive grants	4	3	3	2	4
• Intensive behavioural interventions	2	2	2	2	2
• Sustainable procurement	3	3	3	3	3
Total	52	37	39	32	48

Note: * Number of tools discussed; ** governments can (and often are) involved in more than one role.

In 32 out of these 52 innovative tools governments have, in one way or another, taken up enforcement tasks or seek to ensure compliance with these in other ways (61 per cent). This relatively far-reaching involvement needs to be considered with some care. Generally, governments appear to take up guarding roles in the tools they initiate. Governments appear less keen to take up this role, or are less requested to take up this role, for innovative tools they have not initiated. The sample of 52 collaborations, voluntary programmes and market-driven governance tools did not include any example of a tool that was characterized by an absence of governmental involvement in initiating or leading, and a presence of governmental involvement in enforcing or guarding – because of the nature of the sample it is of course not to say that such examples do not exist.

The broader literature on collaborative governance, voluntary programmes and market-driven governance considers the involvement of governments of significant importance to ensure compliance through enforcement or monitoring of non-mandatory governance tools (see the literature discussed in Chapters 3 and 4). This raises some concerns about the lack of governmental involvement in the enforcement of a number of the examples discussed in this book.

Finally, in almost all of the innovative tools discussed, 48 out of 52 (92 per cent), governments actively supported these tools. Such support comes in the form of funding or administrative support. The small grant that allowed Green Strata (discussed in Chapter 4) to grow out to an internationally acknowledged example of retrofitting strata buildings is a telling example of how little support a group of enthusiastic individuals needs to achieve significant results. The SMART 2020: Cities and Regions Initiative is a telling example that cities may be willing to explore how they can become more sustainable and resilient, but that they sometimes need a push from non-governmental actors to actually take action (discussed in Chapter 4).

5.1.3 Trend 3: Much Innovative Governance for New Commercial Property, Less So for Existing (and) Residential Property

Residential buildings are considered to contribute about as much to global carbon emissions as commercial buildings (at least in developed economies such as Australia; Australian Government, 2008). It is therefore striking to see that the majority of the innovative governance tools discussed in Chapters 3 and 4 have a focus on the commercial sector only (offices, schools, hospitals, retail and so on).

Table 5.2 gives an overview: 24 out of the 52 innovative governance tools discussed predominantly or solely address commercial property (46 per cent). Nine of these are best-of-class benchmarking tools that can be applied to commercial and residential properties alike, but are applied, by and large, to commercial property and hardly to residential property (Cole and Valdebenito, 2013; Fowler and Rauch, 2006a; McGraw Hill, 2011, 2012; Yudelson and Meyer, 2013; see also Chapter 4). Only nine tools (17 per cent) have a sole focus on residential property, while a mere seven tools (13 per cent) seek to address both commercial and residential property. The remaining 12 tools (discussed in Chapters 3 and 4) (23 per cent) seek to indirectly improve sustainability and resilience of commercial and residential property.

Table 5.2 Scope of the innovative governance tools (commercial/residential)

Type	No.*	Scope of the tools discussed			
		Commercial property	Residential property	All property	Other
Collaborative governance	18	3	3	3	9
• Government-led collaborations	10	3	0	1	6
• Private sector-led collaborations	4	0	2	0	2
• Civil society-led collaborations	4	0	1	2	1
Voluntary programmes and market-driven governance	34	21	6	4	3
• Classification and best-of-class benchmarking**	11	9	0	1	1
• Green leasing	2	2	0	0	0
• Private regulation	4	0	2	1	1
• Innovative financing	8	5	1	2	0
• Contests, challenges, competitive grants	4	2	1	0	1
• Intensive behavioural interventions	2	0	2	0	0
• Sustainable procurement	3	3	0	0	0
Total	52	24	9	7	12

Note: * Total number of tools discussed; ** although many best-of-class benchmarking tools are applicable to both commercial and residential property, their uptake is predominantly in the commercial sector (see Chapter 4).

There appears to be a clear logic of why these innovative tools have a greater focus on the commercial sector rather than the residential sector. First, the costs of building a commercial building and the cost savings that can be achieved in the commercial sector outweigh those of the residential sector. In absolute terms the cost savings (that is, energy reduction, water reduction, material reduction, but also insurance premium reduction) are much larger for a large commercial building or a large commercial tenant than for an average residential building or a household. As already discussed in Chapter 4, a 10 per cent cut in energy costs may in absolute terms add up to a considerable amount of money for a large commercial tenant of an office building in the United States, while the same 10 per cent energy cut only adds up to at best US$10 per month for a household (EIA, 2013).

Second, leadership matters in the commercial sector, and especially for owners or tenants of commercial property. For larger firms the sustainability credentials of their building may be a way to showcase their concerns about societal problems, a way to respond to their shareholders' or clients' demands, and a way to meet their social or environmental corporate responsibility strategies (Dixon et al., 2009; Khanna and Anton, 2002). Households do not face such incentives, or face them to a much smaller extent.

Third, the level of professionalism is much larger in the commercial sector than in the residential sector. For property owners and building managers in the commercial sector looking after property is their job; they do so on a day-to-day basis. It is more likely that they are, in one way or the other (for example, through discussions with peers, through ongoing education, through business magazines and conferences), exposed to developments related to urban sustainability and resilience. Homeowners (and owners of other residential property) are very personally involved in their property, and the acquiring, building or retrofitting of a house is most likely a once or at best a few times in their lifetime experience. It is thus unlikely that they know all the ins and outs of the latest developments in urban sustainability and resilience. In addition, professionals in the commercial sector often take risks with their organizations' funds, not their private funds. This may partly take away the deep emotional involvement that homeowners may feel about their property (Gurney, 1999).

It should be noted that even within the commercial sector major differences appear to exist between the demand for buildings with high levels of environmental sustainability from large businesses and those from medium-sized and small businesses. The majority of tools discussed have a strong focus on incentivizing larger businesses and less so on medium-sized and small ones. The Boston-based Sustainable Business Leader Program is one of the few exceptions in the sample that has a specific focus on small and medium-sized businesses (discussed in Chapter 4).

A stronger focus on new than on existing buildings

New buildings are only a marginal contributor to global carbon emissions as compared to existing buildings (IEA, 2009). Both groups of buildings, nevertheless, have a comparable focus in the majority of the innovative governance tools discussed in Chapters 3 and 4.

Table 5.3 gives an overview: 12 out of the 52 innovative tools predominantly or solely address new and future buildings (23 per cent). Seven of these are best-of-class benchmarking tools that can be applied to new and existing buildings, but are applied, by and large, to new and future buildings and hardly to existing buildings (Cole and Valdebenito, 2013; Fowler and Rauch, 2006a; McGraw Hill, 2011, 2012; Yudelson and Meyer, 2013; see also Chapter 4). Another 12 tools have a sole focus on existing buildings (23 per cent), while 22 tools focus on new and future as well as existing buildings (42 per cent). The remaining eight tools discussed in Chapters 3 and 4 (15 per cent) seek to indirectly improve sustainability and resilience of new and existing buildings.

Table 5.3 Scope of the innovative governance tools (new/existing buildings)

Type	No.*	Scope of the tools discussed			
		New/future buildings	Existing buildings	All buildings	Other
Collaborative governance	18	0	5	8	5
• Government-led collaborations	10	0	2	5	3
• Private sector-led collaborations	4	0	2	1	1
• Civil society-led collaborations	4	0	1	2	1
Voluntary programmes and market-driven governance	34	10	7	14	3
• Classification and best-of-class benchmarking**	11	7	0	3	1
• Green leasing	2	0	0	2	0
• Private regulation	4	1	1	1	1
• Innovative financing	8	0	4	4	0
• Contests, challenges, competitive grants	4	2	0	1	1
• Intensive behavioural interventions	2	0	2	0	0
• Sustainable procurement	3	0	0	3	0
Total	52	12	12	22	8

Note: * Total number of tools discussed; ** although many best-of-class benchmarking tools are applicable to both new/future and existing buildings, their uptake is predominantly in new/future buildings (see Chapter 4).

At first glance this is a hopeful picture. New and future as well as existing buildings are addressed to a similar extent in innovative governance tools within the sample discussed. The picture is less rosy when keeping in mind that existing buildings account for the majority of carbon emissions and other wastes produced, as well as the majority of energy and water consumed.

When combining the data from Table 5.2 with that from Table 5.3 it becomes clear that the innovative tools in the sample discussed predominantly focus on new and future commercial buildings, and largely neglect existing residential buildings.

This trend in innovative governance tools is out of sync with how the commercial and residential sectors, and how the clusters new and existing buildings contribute to and are subject to climate risks.

5.1.4 Trend 4: Much Innovative Governance for Urban Sustainability, Less So for Urban Resilience

Another striking trend that comes to the fore when looking closer at the sample of 52 innovative governance tools is that they predominantly seek to achieve urban sustainability and to a much lesser extent urban resilience.

Table 5.4 gives an overview: 40 out of the 52 innovative tools discussed predominantly or solely seek to achieve urban sustainability (77 per cent). Only three have a single focus on urban resilience (6 per cent), while nine have a focus on both urban sustainability and resilience (17 per cent). These numbers are striking. All the more because so many governments, businesses, NGOs and civil society groups call for a focus on urban sustainability *and* resilience; and because the terminology of urban resilience appears to be slowly replacing the terminology of urban sustainability, as discussed in Chapter 1.

One of the reasons for this major difference in focus in the tools discussed may be found in the relatively recent uptake of improved urban resilience to human-made and natural hazards. All tools discussed had been in place for more than two years at the time of study. It is to be expected that there is a time lag between the increased talk about and political and commercial attention for improved urban resilience and the implementation of actual governance tools that seek to achieve improved urban resilience.

Another reason may be that urban resilience is relatively easy to address through statutory regulation for new and future buildings. Construction codes have a long history of addressing building safety (see Chapter 2) and it is likely that it is easier to ramp up these codes than to

Table 5.4 Scope of the innovative governance tools (sustainability/resilience)

Type	No.*	Scope of the tools discussed			
		Urban sustainability	Urban resilience	Both sustainability and resilience	Other
Collaborative governance	18	8	3	7	0
• Government-led collaborations	10	6	1	3	0
• Private sector-led collaborations	4	1	1	2	0
• Civil society-led collaborations	4	1	1	2	0
Voluntary programmes and market-driven governance	34	32	0	2	0
• Classification and best-of-class benchmarking	11	11	0	0	0
• Green leasing	2	2	0	0	0
• Private regulation	4	3	0	1	0
• Innovative financing	8	7	0	1	0
• Contests, challenges, competitive grants	4	4	0	0	0
• Intensive behavioural interventions	2	2	0	0	0
• Sustainable procurement	3	3	0	0	0
Total	52	40	3	9	0

Note: * Total number of tools discussed.

introduce new codes that address urban sustainability (Lee and Yik, 2004). In other words, there may be a lesser need for addressing urban resilience for future and new buildings through innovative governance tools than there is a need to address urban sustainability of new buildings through innovative governance tools.

That said, the lion's share of buildings that are increasingly at risk to human-made and natural hazards are existing buildings. Upgrading their structural safety and their ability to continue operation in a safe manner under a situation of disaster is less cost-effective than upgrading existing buildings to achieve energy and water consumption reductions. After all, energy and water consumption reductions are guaranteed to partly pay back the investments made. For increased urban resilience it is questionable whether the investment will be paid back. Only if disaster strikes, and only if under such a condition a building is better able to maintain functioning than without upgrades made, may conclusions be drawn as to whether the investment was worth the outcome. In other words, the financial incentive to retrofit buildings for increased urban resilience is less, or at least less certain than for increased urban sustainability (see a comparable argument in Urban Green, 2013).

The Australian insurance industry's Building Resilience Rating Tool effectively plays with these financial incentives. It ranks buildings according to how they are expected to withstand climate risks such as increased storms and floods. The better a building is expected to withstand these risks, the higher its rating and the lower its premium to be insured against the consequences of climate risks (discussed in Chapter 4). This tool is highly comparable with best-of-class benchmarking tools, which have proved to be, to a certain extent, effective in improving urban sustainability in the commercial property sector (discussed in Chapter 4).

5.1.5 Trend 5: Much Innovative Governance Seeks to Stimulate the Uptake of Innovative Technologies, Less So to Stimulate Behavioural Change

A final trend that comes to the fore when looking closer at the sample of 52 innovative governance tools is their dominant focus on stimulating the uptake of (innovative) technological solutions to urban sustainability and resilience, and much less on stimulating behavioural change of building occupants.

Table 5.5 gives an overview: 32 out of the 52 innovative tools discussed predominantly or solely seek to achieve improved urban sustainability and resilience through technological solutions (62 per cent). Only four have a single focus on doing so through behavioural change (8 per cent), while nine have a focus on both technological solutions and behavioural change (17 per cent). The remaining seven tools seek improved urban sustainability and resilience in other ways (13 per cent).

It is difficult to find a reason why there is such a large difference in these foci. Much energy and water can be saved and waste production and carbon emissions can be reduced if building occupants begin to use their buildings in a slightly different manner. For instance, as already discussed in Chapter 1, about 10 per cent of energy consumption in households (in developed economies) comes in the form of energy that is

Table 5.5 Scope of the innovative governance tools (technology/behaviour)

Type	No.*	Scope of the tools discussed			
		Uptake of innovative technology	Stimulating behavioural change	Both technology and behavioural change	Other
Collaborative governance	18	8	1	4	5
• Government-led collaborations	10	2	1	3	4
• Private sector-led collaborations	4	3	0	1	0
• Civil society-led collaborations	4	3	0	0	1
Voluntary programmes and market-driven governance	34	24	3	5	2
• Classification and best-of-class benchmarking	11	10	0	1	0
• Green leasing	2	0	0	0	2
• Private regulation	4	1	1	2	0
• Innovative financing	8	8	0	0	0
• Contests, challenges, competitive grants	4	2	2	0	0
• Intensive behavioural interventions	2	0	0	2	0
• Sustainable procurement	3	3	0	0	0
Total	52	32	4	9	7

Note: * Total number of tools discussed.

wasted by having appliances in stand-by mode, while about 55 per cent of energy consumption in (Australian) office buildings is consumed in after office hours, weekends and holidays when not many people are actually using these offices (Greensense, 2013; IEA, 2001).

Such wastes should easily be reduced to zero for households and to a minimum for offices. For instance, if manufacturers make sure that, when put in stand-by mode, their televisions, computer screens, tablets and other devices give visual information that reminds users about the energy waste of stand-by mode, it is likely that they will actually fully switch off their devices (this is a wild guess, but I build on the work of Cialdini, 2009). This may have a spillover effect on switching off other appliances as well. In offices the answer appears even simpler. Given that most office workers nowadays work on computers, it would be all too easy to develop a screensaver that pops up when they end their working day to remind them to fully switch off all appliances, lights, heating and open or close windows according to the season.

5.2 DESIGN PRINCIPLES FOR URBAN SUSTAINABILITY AND RESILIENCE

In summary, when using a broad brush of generalization for the 52 innovative governance tools discussed in Chapters 3 and 4, they may be characterized as relatively conservative and risk averse. Their focus is predominantly on the lucrative high end market for new and future commercial property, where the application of technology to improve urban sustainability has proved to be cost-effective. Their focus is less on the more complicated and the less lucrative residential sector, less on existing buildings, less on seeking to change the behaviour of building occupants and hardly on seeking to improve urban resilience.

Even though this is a rather sobering conclusion, a number of relevant design principles can be distilled from the set of 52 innovative governance tools, as well as from the set of 16 traditional governance tools, that may help to design better governance tools for urban sustainability and resilience. The following 12 principles stand out:

1. Do not shy away from regulating urban sustainability and resilience.
2. Make enforcement and compliance an integral part of the design of governance tools.
3. Take a holistic approach and do not try to tackle urban problems one by one.
4. Individual governance tools have limited impacts; they can mutually reinforce each other in smart governance mixes.
5. Use experience from governing urban sustainability to adequately govern urban resilience, and vice versa.
6. Different groups and individuals respond differently to similar governance tools, this holds most for commercial and residential property owners.
7. Create collaborations and networks that not only talk the talk but incentivize participants to also walk the walk.
8. Include the aspect of learning in the design of innovative governance tools.
9. Reward, champion and create an environment for leadership.

10. While taking the high moral ground on urban sustainability and resilience is laudable, for most parties in the building sector it is (also) about the money.
11. Do not be blinded by (un)successful governance outcomes elsewhere.
12. Be proud of what has been achieved but focus on what has not yet been achieved.

5.2.1 Principle 1: Do Not Shy Away from Regulating Urban Sustainability and Resilience

Although direct regulatory interventions, and in particular statutory regulation, are often critiqued by policy makers, the popular media and academics alike, this approach to governing urban sustainability and resilience has thus far proved to be effective in a wide set of contexts and for a wide range of issues (Baldwin et al., 2011; Levi-Faur, 2011; Parker et al., 2005). There is much to say about regulatory shortfalls and regulatory overburden (Bardach and Kagan, 1982; Sparrow, 2000), but governments have a long history in successfully applying statutory regulation in urban environments (as discussed in Chapter 2).

That said, statutory regulations come with significant problems for regulating urban sustainability, and to a lesser extent for regulating urban resilience. Our knowledge and understanding of environmental sustainability is continuously changing, and technological solutions are often developed faster than regulatory bodies can develop and implement construction codes.

It is an illusion that governments can adequately regulate urban sustainability and resilience if they are not informed by the industry about current and future technological developments and existing regulatory shortfalls. The currently popular application of performance-based regulations with a prescriptive bottom line (see Chapter 2) may be a way out of the lag between ongoing insights into sustainability and resilience, technological change and regulatory responses.

Construction codes predominantly appear to focus on technological solutions and less so on behavioural change. If occupants are not incentivized to use their buildings and infrastructures accordingly, such regulation has a limited impact in improving the environmental and resource sustainability and resilience to hazards of buildings and infrastructures to the extent possible. Through reduced property taxes, for instance, governments may seek to reward households and other building occupants that consume less energy and water than their peer group, while through increased property taxes they may seek to incentivize those that consume more than their peer group to reduce this consumption.

5.2.2 Principle 2: Make Enforcement and Compliance an Integral Part of the Design of Governance Tools

Any governance tool, may it be a traditional or innovative design, however well designed, needs to be complied with to achieve results. How to achieve compliance is therefore as much a critical part of a governance tool as its goal and regulatory structure.

There are a number of questions that need to be answered when designing governance tools. These include:

- Why would those governed not comply with a tool?
- What are the consequences for an individual or a single organization for non-compliance?
- What are the societal consequences if the tool is not complied with en masse?
- What are the enforcement costs to achieve compliance?
- What are the technical or social complications for enforcement?
- Does the organization that administers a governance tool have the necessary capacity and culture to achieve compliance?
- If not, can it rely on self-monitoring by those governed, or is third-party enforcement required?

Throughout the book a number of enforcement strategies have been addressed. The most traditional, and perhaps most widely applied, strategy is a phasing out of enforcement by first checking buildings, infrastructures and other structures' designs against statutory building regulation, and then later their construction in practice, and sometimes the objects themselves when in use.

This strategy may be traced back to the early days of statutory building regulation, which developed in Europe well before the nineteenth century (discussed in Chapter 2). More recently new enforcement strategies have emerged in this area. For instance, in Chapter 2 some attention was paid to the introduction of private sector organizations and individuals for the enforcement of governmental construction codes. It was found that private sector building inspectors may specialize in specific types of buildings and infrastructures that governments find hard to inspect due to a lack of technical expertise in their staff. Yet, it was also found that such privatization may come with severe accountability issues. It was suggested that a 'networked' approach to enforcement, for instance, by bringing in consumer interest groups or insurance companies to the enforcement process, may yield better results.

From Chapters 3 and 4 it may be concluded that innovative approaches to governance such as collaborations, voluntary programmes and market-driven governance tools are often found to fall short in achieving meaningful outcomes since they lack sufficient enforcement. This only stresses the need to make enforcement and compliance an integral part of the design of governance tools.

Because enforcement is such a pivotal aspect of governance for urban sustainability and resilience I shall turn back to it in more depth in the concluding chapter of this book, Chapter 6.

5.2.3 Principle 3: Take a Holistic Approach and Do Not Try to Tackle Urban Problems One by One

In developing and implementing governance tools for sustainability and resilience governments, businesses and civil society groups and individuals appear to seek to tackle urban problems in a one-by-one manner.

For instance, in seeking to improve urban sustainability throughout Europe the European Commission has paid much attention to addressing the energy performance of buildings through the Energy Performance of Buildings Directive (EC, 2010; see Chapter 2). Yet, urban sustainability includes many more aspects than energy performance only. For instance, a building's water consumption, raw material consumption and waste production matter just as well. Such aspects are addressed in other European directives such as the Waste Framework Directive (EC, 2008) and the Water Framework Directive (EC, 2000a), but not necessarily with a sole focus on urban sustainability, or the building sector in particular. By tackling urban problems, one by one, synergies between various aspects of urban sustainability may be overlooked.

The Living Building Challenge is an interesting example of a benchmarking tool that seeks to achieve synergies between various aspects of urban sustainability (International Living Future Institute, 2014; discussed in Chapter 4). For instance, many benchmarking tools such as LEED may be critiqued for being somewhat inconsistent. They allow for the certification of buildings such as (highly sustainable) parking garages, but by doing so defy their own purposes. After all, these buildings do not challenge citizens to change their behaviour for the better, for instance by using their car less (Alter, 2008). The Living Building Challenge builds on broader ideas about urban sustainability, including car-free living.

The recently introduced building codes in Singapore make for another interesting example (BCA, 2008; discussed in Chapter 2). These allow and challenge building designers and developers to mix and match compliance with a series of urban sustainability criteria related to reduced energy, water and raw material consumption and reduced production of carbon emissions and wastes.

The next logical step would to be to develop governance tools not only by keeping a holistic approach to urban sustainability in mind but also seeking synergies between urban sustainability and urban resilience. In Chapter 1 a number of such synergies were discussed.

5.2.4 Principle 4: Individual Governance Tools Have Limited Impacts; They Can Mutually Reinforce Each Other in Smart Governance Mixes

Another logical next step in designing governance tools for urban sustainability and resilience is to actively seek synergies between various tools. When brought together in smart governance mixes, the whole of traditional and innovative tools may be far more than the sum of its parts. Synergies between various tools are actively sought in, for instance, the city of Sydney where the city government is actively involved in the development and implementation of a number of innovative tools, such as Green Strata, CitySwitch Green Office and Environmental Upgrade Agreements (discussed in Chapters 3 and 4). In being involved in such a wide range of tools the city government is able to build bridges between and bring together property owners and tenants in a single forum. In doing so these property owners and tenants may discuss and find solutions to their buildings that they would not have thought of when being distanced from each other.

For city governments that do not have the scale of a global city such as Sydney, the Netherlands may provide a unique example of how synergies between various tools may be achieved. Here, a specific agency that operates at some arm's length from government, the Netherlands Enterprise Agency, is responsible for documenting, studying, developing, supporting and reporting on innovative governance tools. In being involved in, or being aware of, the wide range of innovative tools in the Netherlands the agency is able to build bridges between different tools and seek for synergies among these (Netherlands Enerprise Agency, 2014).

Throughout the book various examples have been discussed of how voluntary best-of-class benchmarking tools are being taken up by governments to ease compliance with statutory regulation. For instance, LEED is being applied as a baseline for other governance tools. This provides synergies between mandatory and voluntary tools when compliance with LEED becomes a criterion for seeing property taxes waived under tax credit and incentive programmes such as the Green Building Tax Credit Program in the state of Maryland, United States, or the Green Building Incentive Program in the county of San Diego, United States (discussed in Chapter 2).

Another example of how a voluntary tool is combined with a mandatory tool is the voluntary National Green Leasing Policy in Australia (discussed in Chapter 4), which eases compliance with the (partly) mandatory Australian NABERS best-of-class benchmarking tool (also discussed in Chapter 4). There are, however, some caveats to indirectly mandating the use of voluntary tools, particularly those that have been developed or are administered outside the realm of governments. Here, it may be questionable whether accountability mechanisms as applied by governments also apply to the voluntary tools they indirectly mandate, and whether the legitimacy of government is at risk for doing so (this issue was further discussed in the conclusion to Chapter 4).

5.2.5 Principle 5: Use Experience from Governing Urban Sustainability to Adequately Govern Urban Resilience, and Vice Versa

One of the trends in governing for urban sustainability and resilience uncovered in this book is that more activity is going on in terms of innovative tools for urban sustainability than for urban resilience (see above). This is somewhat concerning. After all, because cities are increasingly exposed to human-made hazards and climate risks it appears just as important to focus on improved urban sustainability as well as improved urban resilience (UN, 2013a).

In particular, the existing building stock that does not meet current statutory regulation in terms of urban resilience appears vulnerable to human-made hazards and climate risks as a range of examples in this book have shown. In the development and implementation of innovative tools for urban resilience much can be learned from the development and implementation of (successful) innovative tools for urban sustainability. An example is the Australian insurance industry's Building Resilience Rating Tool (discussed in Chapter 4). This tool for improved urban resilience builds on successful best-of-class benchmarking tools for improved urban sustainability.

The innovative and voluntary tools that seek to improve urban sustainability appear to have achieved some successful outcomes, but these are marginal in contrast to the challenge faced by cities and other urban environments. Here, governments, businesses, civil society groups and individuals may draw important lessons from how urban resilience has become deeply integrated in statutory building regulation.

Below are a few questions we may ask from traditional and innovative governance tools for urban resilience when developing their counterparts for urban sustainability.

- How have governments and other governing actors used crises and disasters to ramp up regulatory requirements and other governance tools?
- Which advocacy groups are normally supporting and which are normally opposing such tools?
- What form, structure and content of these tools appears to 'work' best, in what context and why?
- What outcomes have been achieved in terms of enforcing these tools?

5.2.6 Principle 6: Different Groups and Individuals Respond Differently to Similar Governance Tools, This Holds Most for Commercial and Residential Property Owners

All too often governance tools for urban sustainability and resilience do not keep in mind the heterogeneity of the building sector. The industry is unique because of the wide range of large and small businesses involved, the wide range of trades it spans, the larger variety in clientele and services (that is, highly professional property owners to homeowners) and the even larger variety in products it provides (that is, from rather simple townhouses to mega high-rises and extremely sophisticated industrial complexes). In particular, direct regulatory interventions, but to a certain extent the voluntary programmes and market-driven governance tools discussed as well, have a one-size-fits-all approach. In practice, however, different providers, different clients and different products ask for different governance tools.

A good example is the earlier discussed privatization of building code enforcement in Australia and Canada (see Chapter 2). While private sector building code enforcement is accessible to all parties involved in both countries, it appears that particularly professional clients appreciate the involvement of private sector inspectors. They find high levels of expertise and specialism in these inspectors, which perfectly suits the technologically sophisticated buildings for which they are involved. Homeowners, in their turn, are less likely to appreciate the services from these private sector inspectors. Because they are only involved in building regulation-related issues once or a few times in their lives, they prefer a high level of guidance provided by municipal building inspectors (see also Van der Heijden, 2010c, 2011, 2013a). By accepting and including these differences in clientele in their design, better targeted tools can be implemented to meet the needs of these various clients.

Accepting and following up on such differences may help to improve other governance tools as well. For instance, the developers and administrators of best-of-class benchmarking tools often claim that their tools are designed for both the commercial as well as the residential sector. In practice, however, these tools are by and large used in the top end of the commercial property sector only (see above and also Chapter 4).

After more than two decades of experimenting with such best-of-class benchmarking tools it has become clear that the residential sector does not ask the same from these tools as the top end of the commercial sector. It now seems time to draw a relevant lesson: the currently operational best-of-class benchmarking tools are not generally applicable throughout the building sector. More targeted tools for different parts of the sector are necessary if the successes achieved in the top-end of the commercial sector are to be replicated throughout the sector.

Finally, the example of fraud cases with subsidized solar photovoltaic systems and 'rooftop cowboys' in Australia (see Chapter 2) is, again, another illustration that makes clear that in designing a governance tool it is of importance to keep in mind how different parties in the industry respond to these.

A few of the important questions that need to be asked when keeping in mind that different parties respond differently to similar tools include:

- Do they (all) have sufficient technological expertise to understand the tool?
- Do they (all) need the same level of support?
- What type of buildings are various 'clients' of governance involved in?
- What incentive is the best stimulus for which party?

5.2.7 Principle 7: Create Collaborations and Networks That Not Only Talk the Talk but Incentivize Participants to Also Walk the Walk

Collaborative governance for urban sustainability and resilience is gaining increasing popularity (as discussed in Chapter 3). Collaborations, partnerships and networks come, however, with a major risk that needs to be taken into consideration in their design and implementation. Without sufficient criteria to participants' performance in collaborations, partnerships and networks, and without sufficient disciplinary action for a lack of performance of participants, it is unlikely that they will achieve their intended outcomes.

In these situations collaborations, partnerships and networks may then become forums of much discussion but limited action (see Ansell and Gash, 2008). It is, of course, complicated to find the right balance between setting participation criteria on such a level that they do not frighten off prospective participants, while at the same time challenging these to actually take desired action (see Potoski and Prakash, 2009).

The Australian network CitySwitch Green Office is an illustrative example that has sought to find this balance. It requires participants to meet a particular rating under the (partly) mandatory Australian benchmarking tool NABERS. In doing so the developers of the network have overcome a potential risk of 'greenwash' (Lyon and Maxwell, 2006). Yet, in order to not frighten prospective participants off a series of incentives has been introduced by participating (city) governments: participants are financially and administratively supported; meetings and seminars are organized; and the performance of the leading participants is explicitly advocated to the media by these governments.

Another risk of collaborations, partnerships and networks is that they become elite groups that exclude non-members from lessons learned and other advantages (see Kern and Alber, 2010). The 100 Resilient Cities Centennial Challenge, the Better Building Partnership, the C40 Cities Climate Leadership Group, CitySwitch Green Office and the SMART 2020: Cities and Regions Initiative, for instance, all present some generally accessible best practices and case studies on their websites, but it appears that the really important information is 'hidden' in the members-only sections of their websites. This reduces the possibility of policy learning and policy transfer between participants and non-participants of such collaborations and networks.

This problem is all the more relevant when looking at the participants in these collaborations and networks, such as the C40 Cities Climate Leadership Group. These are predominantly the larger cities in the world. How then will smaller cities and towns learn from and have access to the lessons from these collaborations? For instance, Europe has an urban population of about 375 million people. Only 150 million Europeans live in cities larger than 100 000 people. This leaves more than 225 million Europeans in smaller cities and towns, or 60 per cent of Europe's urban population.[4]

This example shows the importance of a shift in focus of these collaborations and networks from addressing the world's largest cities to appreciating the opportunities and challenges in the world's smaller cities and towns. The ICLEI network (discussed in Chapter 3) needs to be credited for indeed including a large number of smaller cities and towns among its participants.

5.2.8 Principle 8: Include the Aspect of Learning in the Design of Innovative Governance Tools

The tools discussed have provided a wide range of insights as to whether they are able, individually or in smart governance mixes, to overcome the three main governance problems presented Chapter 1. These problems were identified as:

- Governments are slow to react to existing problems of urban sustainability and resilience.
- Introducing new legislation and regulation is often meaningless.
- A number of market barriers stand in the way to capitalize on the economic benefits that more sustainable and resilient can cities bring.

The insights and lessons learned from these tools may be referred to as the 'soft outcomes' of these tools, and their value should not be marginalized (Lyon and Maxwell, 2007; Reid and Toffel, 2009; Rogers and Weber, 2010). Most importantly, from analysing the 52 innovative governance tools as well as the 16 more traditional ones it has become clear that it remains questionable whether the range of innovative governance tools has achieved meaningful results in terms of buildings built and retrofitted (see Lee and Yik, 2004; Managan et al., 2012).

A first verdict is that they have not achieved all encompassing results (see Chapter 4). It is true that, for instance, best-of-class benchmarking tools have achieved impressive results in absolute terms. These results, however, fade when contrasted with the overall unsustainability and potential lack in resilience of today's buildings and cities. It is not to say that this implies we should no longer experiment with innovative approaches to governance for urban sustainability and resilience. Some of the design principles above and some of the ones that follow below provide hands-on lessons to improve existing governance tools for urban sustainability and resilience, or provide guidelines for developing and implementing new ones.

What has also become clear, from analysing these 52 innovative and 16 more traditional governance tools, is that the aspect of learning about their performance is of the essence. But, in order to draw lessons from these tools they need to be developed and implemented in such a way that lessons *can* be drawn. In designing and implementing governance tools it is therefore essential to think about drawing and communicating lessons about their performance (Dolowitz and Marsh, 2000; Howlett, 2011; May, 1992; Rose, 2001). It is essential that data is collected, stored and made available to be able to answer questions such as:

- Why was the tool developed and what were the expectations before implementation?
- Why do participants join these tools?
- What is the influence of incentives on the performance of participants?
- What is the actual performance of the buildings built or retrofitted under them?
- To what extent have expectations been achieved?

Throughout the book a range of inspirational examples has been discussed about how learning can be built into the design of policy tools. For instance, in the Netherlands a special agency at some distance from government, the Netherlands Enterprise Agency (see Chapter 3), is responsible for the development, monitoring, evaluation of and reporting on innovative governance tools. By having a central organization for these tasks, lessons about various tools are drawn in a systemized manner, are documented and made available to a wide range of interested parties.

The PEARL network in India (discussed in Chapter 3) is another illustrative example of how lessons about various innovative tools can be collected, by asking similar questions of the experiences with these tools, by documenting them in a single format and by making best practices available to a wide range of interested parties through a highly accessible website.

5.2.9 Principle 9: Reward, Champion and Create an Environment for Leadership

Although the wide set of innovative governance tools may be critiqued for not having achieved sweeping results in terms of buildings built or retrofitted, some particular characteristics appear to recur in the ones that have been, to a certain extent, successful in doing so. One of these aspects relates to the rewarding and championing of leadership by organizations and individuals.

The building sector is a highly competitive sector. A growing group of architects, building designers and developers seek to push the boundaries of what is possible in terms of urban sustainability and resilience, as evidenced by the ever-growing range of journals, magazines and books that laud sustainable and, to a much lesser extent, resilient buildings and city planning (Garvin, 2014; Newman et al., 2009; Wheeler and Beatley, 2009; Yudelson and Meyer, 2013). At the same time, a growing group of property owners and building tenants seek to commission or demand buildings that are leading in urban sustainability and, to a much lesser extent, urban resilience (for example, Eichholtz et al., 2010; Managan et al., 2012; McGraw Hill, 2012; USGBC, 2010).

There is a clear incentive for the latter to seek buildings with high levels of environmental performance. These property owners and tenants may use the credentials of their buildings to market their concern about particular societal issues, and by doing so listen and respond to their clients and shareholders (Dixon et al., 2009; Khanna and Anton, 2002). Innovative governance tools may respond to this need to reward and champion leadership, and a number of the examples discussed highlights approaches for doing so. For instance, CitySwitch Green Office (discussed in Chapter 3) and the Chicago Green Office Challenge (discussed in Chapter 4) have a yearly ceremony during which the best performing participants are awarded a prize. These awarding ceremonies are given extensive media attention. They are considered a key motivator for participants to pursue leading performance (Van der Heijden, 2013d).

Another aspect of tools that have been, to a certain extent, successful is that they have created an environment for individuals or small groups to take action. GreenStrata (discussed in Chapter 3) started as a small group of owners of units in a strata building. They sought to install solar panels on their building, but faced difficulties in getting sufficient other owners of units in their building to agree on the initiative. Unable to find relevant information about this particular topic on the internet this small group of owners sought support from the city of Sydney. The city understood how the problem faced by this small group was representative for possibly a much larger number of strata buildings within its jurisdiction. It supplied this small group of owners with a modest grant and within two years the problem of the group had turned in to an Australian-wide initiative to address strata buildings. The significant role of individuals also becomes clear in examples such as Open Mumbai and Bombay First (both discussed in Chapter 3).

5.2.10 Principle 10: While Taking the High Moral Ground on Urban Sustainability and Resilience is Laudable, for Most Parties in the Building Sector it is (Also) About the Money

The building sector is undeniably a profit-driven sector (see Circo, 2008; Evans et al., 2009). While the evidence is growing that buildings with higher levels of environmental sustainability are no more costly or are even less expensive to design, construct and operate than traditional buildings (IPCC, 2014; WGBC, 2013), these insights are not commonly shared among the actors in the building sector (Managan et al., 2012; WGBC, 2013).

It is therefore not surprising that another characteristic that binds together the tools that have, to a certain extent, been successful is that these have a strong and clear focus on providing positive financial incentives for their participants. The range of best-of-class benchmarking tools discussed in Chapter 4 provides developers, owners and tenants of a building a particular marketing edge (Dixon et al., 2009), and empirical research shows that sustainable office space may yield higher rents and higher selling prices (Eichholtz et al., 2010; GBCA, 2013a). The various forms of innovative financing discussed in Chapter 4 directly address and may partly take away the financial risks related to sustainable buildings that actors in the sector still perceive (Managan et al., 2012). Finally, the contests, challenges and competitive grants discussed in Chapter 4 combine a reduction of financial risks with awarding and championing leadership.

While seeking to achieve an internal motivation of actors in the building sector to improve urban sustainability and resilience is a laudable goal, little should be expected from governance tools that fully build on altruistic motivations in terms of attracting large numbers of participants and their performance (Alexander, 1987; Dawson, 2004; Graff Zivin and Small, 2005; Guagnano, 2001). This may also be an important insight for the development of direct regulatory interventions. By making clear the financial rewards of higher levels of urban sustainability and resilience, new and future regulation by governments may more likely be expected to be accepted by the building sector (see also 'design principle 3', above).

5.2.11 Principle 11: Do Not be Blinded by (Un)successful Governance Outcomes Elsewhere

The 68 examples discussed in Chapters 2 to 4 are, hopefully, of inspiration for the development of similar and different governance tools elsewhere. I have already stressed that these examples provide but a glimpse of the wide range of governance tools around the globe and the challenges and opportunities they may provide. When drawing lessons from these and other examples those involved in the development and implementation of governance tools should not be blinded by successful or unsuccessful outcomes from elsewhere. Governance outcomes such as those presented throughout this book are highly context specific (Holley et al., 2012; Van der Heijden, 2012; Wurzel et al., 2013).

Typical examples are the various tools that seek to promote the uptake of ESCOs presented in this book (discussed in Chapter 4). Whereas ESCOs have been applied with considerable success in the United States (Goldman et al., 2005; Vine, 2005) and are one of Singapore's key approaches to make a transition to a highly sustainable urban building stock (Briomedia Green, 2012; see Chapter 4), the application of ESCOs faces considerable constraints in the Netherlands (City of Rotterdam, 2011; Hofman, 2013; Simons, 2013).

In the Netherlands the development of the first ESCO contract required considerable resources. Various involved parties were not familiar with ESCO contracting and the Dutch legal framework did not provide for enough safeguards. Supported by the Dutch government the first ESCO contract was developed between 2009 and 2011, and the first ESCO was in operation in 2011. Because of the lengthy and costly development process of this ESCO contract there now appears to be an understanding in the Netherlands that ESCOs more generally are a complicated and risky approach to achieve energy reductions. The participants in this first ESCO, however, point to their contract as providing an example for other parties to build on, which may save them the effort needed to develop an ESCO contract (see Chapter 4).

5.2.12 Principle 12: Be Proud of What has Been Achieved but Focus on What has Not Yet Been Achieved

Without question the various direct regulatory interventions, collaborations and market-driven governance tools have started a transition towards a more sustainable and resilient urban environment. Those involved in current and past governance for urban sustainability and resilience may and should be proud of their achievements. Still, care should be taken to not be blinded by the relatively small successes achieved thus far.

The example of the most 'successful' market-driven governance tool in the building sector is telling. The success of LEED-Platinum (considered the pinnacle of sustainable buildings; see further Yudelson and Meyer,

2013; see also Chapter 4 of this book) in the United States is a mere 0.18 per cent of all built-up space in this country over a period of 20 years. This implies that it takes a growth rate of a factor of more than 500 (or 500 LEED-Platinum copy-cat tools, or an immense uptake of the construction and retrofitting of buildings outside the scope of voluntary programmes and market-driven governance tools) to see the full building stock of the United States having high levels of environmental sustainability by 2034 – levels that are considered to be leading, achievable and cost-effective in 2014 (but see critiques to LEED discussed in Chapter 4).

The focus of our future activities in this area should be to draw relevant lessons from success stories so that we can more systematically address the challenges provided by buildings and cities.

6. Conclusion: in search of an answer to the key question

This leaves me with reflecting on the key question that drove the research presented in the book:

What governance approaches and tools may help to improve the resource sustainability of our buildings and cities, may help to reduce their negative impacts on the natural environment and may make them more resilient to man-made and natural hazards?

Unfortunately, after bringing together, mapping, describing and, to the extent possible, analysing 68 governance tools for urban sustainability and resilience, I cannot provide a single and rosy answer to this question. The three main governance problems discussed in Chapter 1 have not yet been overcome by the traditional and innovative governance tools discussed. These problems were identified as:

- Governments are slow to react to existing problems of urban sustainability and resilience. It often takes a long time to develop and implement legislation and regulation, and even longer for these to cause their effects.
- Introducing new legislation and regulation is often inconsequential. In developed economies cities develop too slowly for new legislation and regulation to be meaningful. In developing economies cities develop too rapidly for new legislation and regulation to be meaningful.
- A number of market barriers stand in the way to capitalize the economic benefits that more sustainable and resilient can cities bring.

Traditional approaches to governance, such as direct regulatory interventions led by governments, such as statutory regulation, subsidies and other economic incentives, still make up the lion's share in addressing urban sustainability and resilience; and collaborations, voluntary programmes and market-driven governance in this area are still very small pockets of good practice. With traditional approaches to governance still being so dominant, two of the three main governance problems are not yet sufficiently addressed.

The existing building stock remains largely outside the scope of future statutory regulation. This implies that around the globe about 98 per cent of buildings and infrastructures will, normally, not have to comply with new statutory regulation by governments. It also remains questionable whether novel direct regulatory interventions are able to respond in a timely fashion to the rapidly growing urban problems in rapidly developing economies. This is mainly a problem that comes with the enforcement of statutory building regulation.

Related to these two points, there is a tendency in the building sector to develop and implement one-size-fits-all solutions, while the problems faced and the sector itself show a high level of complexity and variety. This holds for the traditional as well as the innovative governance tools discussed throughout the book.

Finally, even within the small pockets of good practice created by collaborations and market-driven governance tools the outcomes are all but rosy. These non-traditional approaches to governance predominantly focus on the top end of the commercial property market, they aim to increase urban sustainability and they seek to do so through an increased uptake of innovative technologies. The residential property market, the existing building stock more generally, urban resilience and the application of social know-how about behavioural change is largely outside the scope of these innovative collaborations and market-driven governance tools.

In other words, when looking outward from within these small pockets of good practice it appears that a lot of activity is going on. Yet, when looking into these pockets from the larger problems faced it appears that this activity is going on in a very small part of the building sector.

What then is needed?

6.1 A BRAVE APPLICATION OF STATUTORY REGULATION IS REQUIRED

An important part of the way forward, in my view, is the implementation of more ambitious, and more targeted statutory regulations that mandate the use of the cost-effective technologies and behavioural change that are currently available (see Section 6.3 for the idea of 'targeted regulation', and how such targeted regulation may overcome the current shortfalls of direct regulatory interventions in achieving urban sustainability and resilience).

I understand that this goes against modern libertarian thinking about regulation, and against an ongoing policy mantra of deregulation. But if the opportunities of the building sector to become more environmentally and resource sustainable and resilient to human-made and natural hazards are to be unleashed with increasing speed, then targeted regulation is necessary.

Over the last decades the building sector has not shown to be willing to voluntarily make a significant shift towards higher levels of sustainability or resilience, even though the evidence has for long been available that this shift can actually be very profitable. It is, I think, unlikely that this complicated industry, with so many vested interests and so much capital invested in existing property and other assets will suddenly start moving in the right direction without a push.

In addition, governments need to be brave and require mandatory retrofits of the highly unsustainable and sometimes highly non-resilient existing building stock. It is an illusion that the potential of carbon emission reductions, reduced energy, water and material consumption, and reduced wastes in cities will be achieved if the existing building stock and existing infrastructures are not upgraded. This has been stressed, time and again, by governments, businesses and NGOs alike (for example, Deloitte, 2013; EC, 2013; IEA, 2009; Urban Green, 2013).

At the same time, increased sharing of knowledge and resources appears necessary from developed to (rapidly) developing economies. Problems with, for instance, performance-based construction codes, private sector building regulatory enforcement and subsidies for the application of renewable energy in urban settings have been experienced and have been addressed repeatedly by governments in developed economies. They now have a reasonable insight into what type of intervention may be effective in addressing a problem and in what context. Even though it is unlikely that such knowledge is directly applicable in the context of (rapidly) developing economies, the latter may leap-frog over the governance problems faced earlier by governments in developed economies. Networks such as ICLEI, the Mexico City Pact and the C40 Cities (discussed in Chapter 3) are in place for a rapid sharing of this knowledge.

That said, governments in rapidly developing economies need to act to reported shortfalls in the enforcement of construction codes, and a reported unwillingness of taking disciplinary actions when non-compliance is found (Hettige et al., 1996; Kirkpatrick and Parkers, 2004;

Nath and Behera, 2011). It is therefore hopeful to see that governments in, for instance, Malaysia and India now have policies and agencies in place that exactly address this issue (Siddiquee, 2010; Transparency International, 2009).

Yet, ramping up existing regulation in the area of urban sustainability and resilience, introducing new ones and mandating retrofits is unlikely to be a politically attractive move both in developed and (rapidly) developing economies (Lee and Yik, 2004).

After decades of deregulation and cutting red tape (for good discussions, see Jordana and Levi-Faur, 2004; Vogel, 1996) a politician will think twice before she proposes introducing more and stricter regulation. She will, however, think more than twice about doing so if she is also aware of her electorate. Introducing more stringent building regulation and mandatory retrofits directly affects almost all of a politician's electorate. It affects all home or other property owners; it affects employees in the building sector; it affects immensely powerful lobby groups that also provide many policy makers with necessary financial support; and on top of all that, it affects a sector that globally adds to more than 11 per cent of economic activity (GCF, 2009; Raman and Shove, 2000).

Thus, even a brave politician is unlikely to see her proposal for increased statutory building regulation implemented because proposing this idea is likely to be the end of her political career – she will lose votes and financial support. In short, while I conclude that a brave application of statutory regulation is required, I do not expect that policy makers will find it easy to follow this insight.

6.2 NETWORKED ENFORCEMENT IS NEEDED

Further, just introducing more statutory regulation will unlikely be effective if enforcement remains unchanged.

Throughout the book it has become clear that the currently dominant approach to the enforcement of statutory building regulation comes with many shortfalls. As highlighted in Chapter 2, time and again inquiries to fatal building incidents conclude that insufficient enforcement was carried out by the responsible authority or the organization that was delegated this task (for example, municipal building inspectorates, private sector inspectors, self-regulated businesses).

Two dominant problems recur. First, often those responsible for enforcement did not take sufficient action to enforce regulation because they were, in one way or the other, incentivized for not doing so. Second, those responsible for enforcement did not take sufficient action because they were not technically capable of carrying out adequate monitoring and enforcement.

The building sector is known for corruption, bribery and fraudulent activities. Such activities are reported in building regulatory regimes in which governments carry out enforcement, in those in which such enforcement is privatized, and in systems where enforcement is largely carried by those subject to regulations themselves (for example, Broeders and Hakfoort, 1999; Davis, 2007; Hunn, 2002; Nwabuzor, 2005; Sohail and Cavill, 2008; The Japan Times, 2005; The Yomiuri Shimbun, 2005; Yates, 2003). Also, such activities are found in various segments of the building sector, from the development of houses to the construction of multi-million dollar office buildings. In other words, it is likely that the norms in the building sector complicate and limit the effectiveness of building regulatory enforcement regimes. Changing the actors who carry out these inspections does not appear to be a sufficient solution to this issue.

Building on a few promising examples from my own studies, and on larger evidence from studies in other areas (for example, Baldwin and Black, 2008; Braithwaite, 2004; Drahos, 2004, 2013; Grabosky, 2013), I argue that the traditional approach to the enforcement of statutory regulation in the building sector is outdated and that a new approach, networked enforcement, is needed. The traditional approach is a regime where the actor subject to statutory building regulation (for example, a developer, contractor, trades person, property owner) is accountable to only one other actor for complying with statutory building regulation (for example, a municipal building inspectorate or a private sector inspector).

This traditional approach puts too much responsibility and too much disciplinary power in the hands of the actor responsible for enforcement. If this actor, intentionally or not, carries out insufficient monitoring, or does not respond adequately to violations by taking no or limited disciplinary actions, then violations may go unnoticed.

I expect more from networked approaches to the enforcement of building regulation. In such a networked enforcement regime the actor subject to statutory building regulation is accountable to a range of other actors, and these all have specific disciplinary powers to incentivize compliance. Figure 6.1 provides a schematic illustration of a promising example from one of my own studies, certified professionals in Vancouver, Canada (Van der Heijden, 2009, 2010b, 2010c).

To overcome a shortage of staffing, both quantitative and qualitative, in the city of Vancouver's municipal building inspectorate in the 1980s a system of certified private sector building inspectors was introduced.

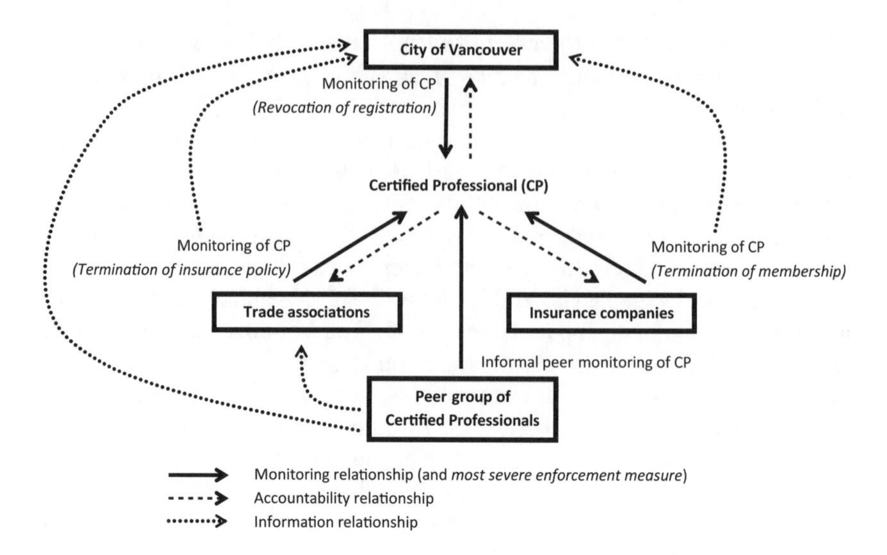

Figure 6.1 Schematic representation of networked building regulatory enforcement regime

Under the Vancouver building regulatory regime developers and property owners can choose to involve a municipal enforcement officer or a private sector certified professional in their construction projects. To ensure accountability of this building regulatory enforcement regime these certified professionals are subject to a networked enforcement regime in which the municipality, insurance companies and trade associations participate.

In order to become a certified professional, an individual needs to be acknowledged and registered by the city. Requirements to registration are related to the prospective certified professional's level of education, working experience and ongoing professional development. In addition, the prospective certified professional needs to hold a particular level of professional indemnity insurance, and needs to be a member of the architects' association or engineers' association. Once registered, the certified professional is accountable to all three organizations, and all three organizations have their own disciplinary measures to incentivize compliance.

First, the municipality can inspect the work of the certified professional and if violations are found it can, ultimately, withdraw the certified professional's registration, which *de facto* implies that the certified professional can no longer work in this profession. The city, however, applies a responsive regulation enforcement strategy (discussed in Chapter 2) and seeks compliance first through education and consultation, and if the certified professional does not respond to these forms of enforcement, then the city may issue a fine or eventually revoke the certified professional's registration.

Second, the trade associations may terminate the certified professional's membership, if for some reason it is found that the certified professional does not provide work at an acceptable level of competence intended. The associations have in place, for instance, a system to investigate consumer complaints made to certified professionals. If they decide to terminate a certified professional's membership, then the certified professional also loses the necessary registration with the city. These associations have a strong incentive to ensure compliant members. After all, one rogue member may change the general public's view of the association for the worse.

Third, an insurance provider may decide to no longer provide a certified professional with a professional indemnity insurance policy. This can, for instance, be done if the insurance company has to pay out too often for works the certified professional has completed. Again, this puts the certified professional out of business because without necessary insurance the certified professional cannot be registered with the municipality. It goes without saying that insurance companies have a strong financial incentive to terminate insurance policies of rogue certified professionals.

In addition, certified professionals are subject to a strong form of informal peer monitoring. Because the group of certified professionals in Vancouver is relatively small the certified professionals know or hear of each other's work. They themselves have a strong incentive to make rogue certified professional known to their trade association or the municipality because one rogue certified professional may change the general public's opinion about their profession, or because insurance companies may decide to increase overall insurance premiums when they have to pay out too many certified professional-related claims.

In short, in this example an individual certified professional is subject to three organizations that all have 'big sticks' that they not only threaten to use but can use with relative ease (Braithwaite, 2004). The consequences of either one of these organizations using their 'sticks' is the same: it will put the certified professional out of work.

These organizations can also easily tip each other when they suspect a certified professional is not abiding by the rules but do not have the capacity or power to carry out enforcement or take disciplinary action. In particular, the trade associations and the insurance companies have strong self-interested motivations to make such knowledge known to the city of Vancouver. This networked enforcement regime is strengthened because of the role the other certified professionals play as highly effective and self-interested 'surrogate enforcers' (Gunningham and Grabosky, 1998).

This particular networked enforcement regime shows much similarity with the collaborative governance examples discussed throughout this book; and such collaborations may very well be a model for future networked enforcement regimes.

There appears no shortage of network partners in the building sector for governments to collaborate with in networked building regulatory enforcement regimes. Insurance companies, banks and other finance providers, trade associations, educational facilities, consumer representative groups, ombuds(wo)men, practitioner organizations and individuals in the sector, citizens and NGOs may be included in networked building regulatory enforcement regimes.

Of course, it remains questionable whether such a regime could have prevented the collapse of the Ranza Plaza building in Dhaka, Bangladesh (see Chapter 2). Yet, developments in the aftermath of this incident at least show willingness of other actors to be involved, directly or indirectly, in the enforcement of statutory building regulation. The collapse has attracted international media attention (for example, *The Economist*, 2013), which in turn has put the safety of Bangladesh factory workers in the spotlight of advocacy groups and NGOs such as Human Right Watch and the Industrial Global Union. These NGOs further spurred media attention in the incident and asked, among others, the (Western) fashion industry to take responsibility. Even before the incident some brands were moving their production to other countries because of dangerous working conditions and unsafe factories in Bangladesh (*Wall Street Journal*, 2013). The fashion industry has further responded by setting up the Ranza Plaza Donors Trust Fund to compensate victims of the collapse (*New York Times*, 2013).

For the authorities responsible for building regulation and its enforcement in Bangladesh, but also in other rapidly developing economies, this appears a window of opportunity to include these various organizations in networked enforcement regimes.

6.3 THE COMPLEXITY OF URBAN GOVERNANCE ASKS FOR TARGETED STATUTORY REGULATION AND OTHER GOVERNANCE TOOLS

Over the last ten years I have had the opportunity to study, in particular, the global transition towards buildings with higher levels of environmental performance than traditional buildings. I have had the opportunity to discuss this topic in interviews with hundreds of policy makers, administrators, developers, financiers, building sector practitioners of all sorts and academics. In addition, I have read much of the academic, policy and practitioner literature on the topic in the field, and the extensive list of references to this book is but a small part of that literature.

Looking back at these ten years of research, the interviews held and the literature read one thing strikes me as remarkable. When I discuss and read about, for instance, the transport sector, there hardly seems to be any discussion about whether the subsector of aviation should be governed differently from the subsectors of road transport, shipping or train transport. Within these subsectors, say road transport, again clear distinction is made in transport by truck, car, motorbike, bicycle and foot. We do not expect that a policy maker with expertise in aviation also has in-depth knowledge of shipping, and train transport and road transport. Even more, we would likely feel very uncomfortable if, say, the city of New York's inspectors of roadworthiness of private vehicles (a statutory requirement) would also be responsible for checking whether the airplanes that land at and depart from the city's airports meet statutory regulation.

Yet, for the building sector we expect exactly this.

Even though the building sector shows a broad variety of building (and infrastructure) types, these are often all governed by a small group of policy makers within a ministerial department, and much governing of the building sector is delegated to local authorities. Such local building and infrastructure authorities often have a very limited budget for staff, which results in the remarkable situation that a handful of people are responsible for regulating and enforcing the broad variety of possible building types. From sheds to one-storey single family houses, from hospitals to shopping malls, from high-rise condominiums to multimillion dollar office buildings, from car workshops to power plants. Society appears to have become so used to thinking that buildings are 'just buildings' that the current variety and complexity of the sector, and related the variety and complexity of governance for urban sustainability and resilience is hardly questioned.

This book has given some insight to this variety and complexity of urban governance, if only for the building sector. It has shown that some governance tools may be effective for particular parts of the building sector, for instance, the top end of the commercial market; but not for other parts, for instance, the market for small and medium-sized businesses. In a similar vein, it has shown that businesses and civil society groups and individuals may yield hopeful outcomes in governing certain parts of the building sector, for instance, in universities and other educational facilities; but not in other parts, for instance, the residential sector.

A one-size-fits-all approach to governing urban sustainability and resilience is unlikely to be an adequate response to the variety and complexity of the sector described. Only by embracing and acting to this variety may a meaningful transition to urban sustainability and resilience be achieved. This holds for traditional and innovative governance tools alike. In particular, for traditional governance tools such as statutory regulation such an embracing of the complexity of the sector may result in a situation where this complexity is first untangled and can then be addressed with 'simple rules' (Epstein, 1995).

When developing targeted regulations and governance tools that do not seek one-size-fit-all solutions for the sector as a whole, but that target a specific problem in a specific subsector, and that address a specific group of actors subject to regulation, more regulatory clarity may be achieved. The wide range of collaborative tools, voluntary programmes, and market-based governance tools discussed highlight the potential of targeted regulations and governance tools.

In various countries' building regulatory regimes discussed in this book a first step has already been made in that direction. For instance, the Building Codes of Australia distinguish between different classes of buildings and different sets of statutory regulations apply to these (ABCB, 2011). Relatively easy buildings such as houses, sheds and carports (referred to as class 1 and class 10 buildings) are subject to their own set of regulations. These take into account that such buildings are often owned and maintained but are also often developed and constructed by owner-builders. They have fully different technological knowledge than professionals in the industry, and thus ask for a different regulatory approach.

By clearly distinguishing in the 'clients of governance', owner-builders are not overly burdened with regulations that do not apply to them. In a similar vein, professionals in the industry are not overburdened with regulations that do not apply to them.

This should, in my opinion, be the goal of still current deregulation attempts. It is not about ensuring an overall, quantified, limited number of regulations but about ensuring a limited and targeted set of regulations that apply to specific parts of the building sector and the actors in it. More targeted regulation may be quicker to develop, easier to apply and understand, and simpler to enforce by inspectors who are trained to enforce particular parts of the sector. Here much can be learned from the wide range of innovative governance tools discussed in this book.

6.4 IT IS ESSENTIAL FOR GOVERNMENT, BUSINESS AND CIVIL SOCIETY TO TAKE UP NEW GOVERNING ROLES

Technology-oriented sciences have over the last decades evidenced the opportunities that new and traditional technologies have to offer to improve the environmental and resource sustainability of buildings and cities and their resilience to human-made hazards and climate risks. Behaviour-oriented sciences have evidenced the opportunities that behavioural change on individual and societal levels has to offer in this area. Governance-oriented sciences have recently begun to evidence what collaborations of governments, businesses and civil society groups and individuals, and market-based governance tools have to offer to the transition towards meaningful urban sustainability and resilience.

Over recent decades governments, businesses and civil society groups and individuals have begun to take up new roles and responsibilities in governance for urban sustainability and resilience. While still being the dominant authority in developing and implementing statutory regulation in this area, governments are now also actively involved in, bring together, collect data on and distribute insights about innovative governance tools. They do so, as discussed throughout the book, to incentivize businesses and citizens as well as (other) governments to voluntarily construct buildings with higher levels of environmental and resource sustainability and that are better able to withstand hazards than that required by statutory regulation or to voluntary retrofit and upgrade existing buildings.

Businesses, in turn, are actively participating in the development and implementation of such tools, and appear attracted to joining them as subjects. They do so to push the boundaries of what is possible in terms of sustainable and resilient buildings and cities, to see their leadership acknowledged, to steer policy agendas, to improve their reputation and to simply make a profit (Hirokawa, 2009; McGraw Hill, 2011).

Civil society groups and individuals actively participate in the development and implementation of these tools, and join these as subjects. They do so for comparable reasons as businesses, but also to see addressed the particular public interests for which they advocate. The classical example is the 1999 campaign of Greenpeace against Home Depot (then the largest supplier of do-it-yourself products in the United States), which resulted in Home Depot seeking certification for all its timber products from the Forest Stewardship Council (Domask, 2003).

These new roles taken up by governments, businesses, civil society groups and individuals are key in a meaningful transition towards cities and other urban environments that are less dependent on energy, water and other resources, produce fewer greenhouse gases and other wastes, and can better withstand human-made hazards and climate risks. They appear of more importance than the particular structure of the governance tools implemented that seek this transition.

After all, when unpacking these tools they all look more or less the same: a set of regulations is introduced to describe a particular goal that is sought and the approach for reaching it is to a certain extent spelled out; a system of monitoring is introduced to measure compliance with these regulations and thus to gain insight into the extent to which the goal is or is likely to be achieved; and a set of rewards and disciplinary actions is crafted to ensure compliance with these regulations.

Another important part of the way forward, I therefore argue, is to better understand *what* new roles are taken up by *which* parties in the different approaches to governance for urban sustainable resilience discussed throughout Chapters 2 to 4; and, *how* these parties and their (new) roles exactly contribute to reduced consumption of energy, water and resources in cities, to reduced greenhouse gas emissions and production of other wastes in cities, and to cities that can better withstand human-made hazards and climate risks.

Through this book I have sought to take a small step along that way by bringing together, mapping, discussing and, to the extent possible, analysing a large series of governance tools for urban sustainability and resilience, and the new roles of governments, businesses and civil society in them.

Appendix: methods

The data in this book originates from a series of research projects that I carried out between 2005 and 2013.

All projects sought to better understand the governance of urban affairs such as urban sustainability and urban resilience (Van der Heijden, 2013a, 2013b, 2013d, forthcoming 2014a, forthcoming 2014b, forthcoming 2014c), energy performance of buildings in Europe (Van der Heijden and Van Bueren, 2013; Vermande and Van der Heijden, 2011)[1] and compliance with building codes (Van der Heijden, 2009, 2010a, 2010b, 2013a). These projects built on data collected in Australia, Canada, Germany, India, Malaysia, the Netherlands, Singapore, the United Kingdom and the United States.

In all these research projects I relied on a small to medium-sized sample of cases to study (10 to 50 cases per project). In all research projects these cases, or units of analysis, are individual governance tools. For instance, the regulatory regime of private sector building code enforcement in Vancouver (Van der Heijden, 2010b); the governance tool that has emerged as a result of the European Energy Performance of Buildings Directive in the Netherlands (Van der Heijden and Van Bueren, 2013); or the voluntary best-of-class benchmarking tool Green Star in Australia (Van der Heijden, 2013d).

The advantage of a small to medium-sized study is that it enables in-depth engagement with the complex nature of the development and implementation of governance tools. It helps to capture and understand their context, structure and the actors involved (in my approach to research I am mostly inspired by Brady and Collier, 2004; Dunn, 2003; King et al., 1994; Mahoney and Goertz, 2006; Rihoux and Ragin, 2009; Sabatier, 2007).

A.1 CRITERIA FOR SELECTING THE CASES FOR THIS BOOK

I sought to provide, discuss and analyse to the extent possible a broad overview of traditional and innovative governance tools for urban sustainability and resilience, based on my earlier studies.

The book predominantly builds on a large set of cases that I have analysed in the various studies mentioned before. I have selected cases from this set following a number of criteria. These criteria are derived from the governance literature discussed throughout the book. The brief theoretical introductions to each chapter best capture these for each of the cluster of cases, but a number of general selection criteria can be derived from these.

First, in Chapter 1 governance was defined as an intended activity undertaken by one or more actors seeking to shape, regulate or attempt to control human behaviour in order to achieve a desired collective end.

The first selection criteria for the cases discussed was their focus on urban sustainability and resilience. Each case presented builds on a set of rules that states an ambition, goal or desired state of affairs. It is further specified how this is to be achieved, and what incentives are in place for achieving compliance with these rules. For instance, Green Star (discussed in Chapter 4) seeks to, among others, 'establish a common language [for sustainable development in the property industry, and] set a standard of measurement for built environment sustainability' (GBCA, 2013b). It has in place an extensive set of codes and regulations that specify how participants in the arrangements can meet a certain Green Star performance, and it awards their buildings with a particular rating according to their performance.

Second, in order to be able to map, describe and analyse, to the extent possible, the range of traditional and innovative governance tools that are currently applied throughout the world I have selected cases to illustrate variety in the structural conditions of these tools. These structural conditions are, among others:

- the type of regulatory requirements (i.e., prescriptive, performance or goal-based regulations)
- the approach to monitoring (i.e., self-monitoring, government monitoring, third-party monitoring)
- the approach to enforcement (i.e., coercion, support)
- the incentives in place for (non-)compliant behaviour (i.e., fines, subsidies, information, certification).

Third, through this book I seek to gain insight into the opportunities and constraints governments and NGOs face in governing urban sustainability and resilience around the globe. To meet this goal I have, to the extent possible, included a variety of examples for each of these approaches to governance on international, supra-national, national, regional and local levels. On local, regional and national levels I have included examples from the Asia-Pacific, West Europe and North America. This leaves out a focus on important areas such as Africa, Central America and the Caribbean, East Europe and the former Soviet Union, the Middle East and South America. Further, on the national level the current study lacks insight in dominant countries in Asia such as China (for a good primer on the state of affairs in dominant countries in Asia, see Hong and Laurenzi, 2007) and the oft-lauded Scandinavian countries (for a good overview, see Evans et al., 2005).

Fourth, the main focus of the book is on innovative governance tools for urban sustainability and resilience (Chapters 3 and 4). For these innovative tools I have selected cases to include various foci. For instance, the various roles governments take up in these tools (that is, as initiator, assembler, guardian or supporter); their focus on commercial buildings, residential buildings, or both; on new or existing buildings, or both; on sustainability, resilience, or both; and their focus on technological innovation, behavioural change, or both. Table A.1 gives the coding of the various collaborative governance tools discussed in this book (Chapter 3), and Table A.2 gives the coding of the various voluntary programmes and market-driven governance tools discussed (Chapter 4).

A.2 DATA UNDERLYING THIS BOOK

My main source of data were semi-structured interviews with elite informants such as policy makers, administrators, architects, engineers, developers, financiers and users of buildings. For the various projects, I have targeted particular interviewees for their specific role in a case under scrutiny (for example, a key administrator, key end-user, key supporter or key critic) and been open to referrals to other interviewees by my initial interviewees. This may be understood as targeted snowball sampling (on snowball sampling, see Longhurst, 2003). To address, to the extent possible, the bias of interviewees who are overly enthusiastic about their 'own' case I have sought to include a wide range of elite actors in the pool of interviewees and, where possible, discussed different cases with individual interviewees (Sanderson, 2002). To improve the validity of my interview data I have built in checks and balances in the various interview-questionnaires I have used (Richards, 1996; Silverman, 2001).

Table A.1 Coding of the collaborative governance tools

Name	Type	Role of government				Focus A				Focus B				Focus C				Focus D			
		G1	G2	G3	G4	A1	A2	A3	A4	B1	B2	B3	B4	C1	C2	C3	C4	D1	D2	D3	D4
100 Resilient Cities Centennial Challenge	Civil society-led collaboration	–	–	–	–	–	–	–	X	–	–	–	X	–	X	–	–	–	–	–	X
Better Building Partnerships	Government-led collaboration	X	X	X	X	X	–	–	–	–	X	–	–	X	–	–	–	–	–	X	–
Bombay First	Private sector-led collaboration	X	X	X	X	–	–	–	X	–	–	X	X	–	–	X	–	–	–	X	–
Building Resilience Rating Tool	Private sector-led collaboration	–	–	–	–	–	X	–	–	–	X	–	–	–	X	–	–	X	–	–	–
C40 Cities Climate Leadership Group	Government-led collaboration	X	X	X	X	–	–	–	X	–	–	–	X	–	–	X	–	–	–	–	X
CitySwitch Green Office	Government-led collaboration	X	X	X	X	X	–	–	–	–	–	X	–	X	–	–	–	–	X	–	–
Climate Change Sector Agreements	Government-led collaboration	X	X	X	X	X	–	–	–	–	–	X	–	X	–	–	–	–	–	X	–
Common Carbon Metric	Civil society-led collaboration	–	–	–	X	–	–	X	–	–	–	X	–	X	–	–	–	X	–	–	–
Future Cities Network	Government-led collaboration	X	X	X	X	–	–	–	X	–	X	–	–	–	X	–	–	X	–	–	–
Green Deals	Government-led collaboration	X	X	X	X	–	–	X	–	–	–	X	–	X	–	–	–	X	–	–	–
Green Strata	Private sector-led collaboration	–	–	–	X	–	X	–	–	–	X	–	–	X	–	–	–	X	–	–	–

Growing a Green Heart Together	Government-led collaboration	X	X	X	X	–	–	–	X	–	–	–	X	X	–	–	–	–	–	X	–
ICLEI – Local Governments for Sustainability	Government-led collaboration	X	X	X	X	–	–	–	X	–	–	–	X	–	–	X	–	–	–	–	X
Ontario Regional Adaptation Collaborative	Government-led collaboration	X	X	X	X	–	–	–	X	–	–	X	–	–	–	X	–	–	–	–	X
Open Mumbai	Civil society-led collaboration	–	X	–	X	–	–	X	–	–	–	X	–	–	–	X	–	X	–	–	–
Peer Experience and Reflective Learning Network	Government-led collaboration	X	X	X	X	–	–	–	X	–	–	X	–	X	–	–	–	–	–	–	X
SMART 2020: Cities and Regions Initiative	Private sector-led collaboration	–	–	–	X	–	–	–	X	–	–	–	X	–	–	X	–	X	–	–	–
Sustainable Backyard Program	Civil society-led collaboration	X	X	X	X	–	X	–	–	–	X	–	–	–	–	X	–	X	–	–	–

Note: Abbreviations used are G1 = government in an initiating or leading role; G2 = government in an assembling role; G3 = government in a guarding role. Focus A = what type of buildings does the tool address predominantly? A1 = commercial buildings; A2 = residential buildings; A3 = both commercial and residential buildings; A4 = other focus. Focus B = what type of buildings does the tool address predominantly? B1 = new buildings; B2 = existing buildings; B3 = both new and existing building; B4 = other focus. Focus C = what is the dominant focus of the tool? C1 = urban sustainability; C2 = urban resilience; C3 = both urban sustainability and resilience; C4 = other focus. Focus D = what is the dominant approach to seeking the tool's goal achievement? D1 = through technology; D2 = through behavioural change; D3 = through both technology and behavioural change; D4 = other focus.

Table A.2 Coding of the voluntary programmes and market-driven governance tools

Name	Type	Role of government				Focus A				Focus B				Focus C				Focus D			
		G1	G2	G3	G4	A1	A2	A3	A4	B1	B2	B3	B4	C1	C2	C3	C4	D1	D2	D3	D4
1200 Buildings	Innovative financing	X	X	X	X	X	–	–	–	–	X	–	–	X	–	–	–	X	–	–	–
Aldinga Arts Eco Village	Private regulation	–	–	–	–	–	X	–	–	X	–	–	–	X	–	–	–	–	–	X	–
Amsterdam Investment Fund	Innovative financing	X	X	X	X	–	–	X	–	–	–	X	–	X	–	–	–	X	–	–	–
Billion Dollar Green Challenge	Innovative financing	–	–	–	X	X	–	–	–	–	–	X	–	X	–	–	–	X	–	–	–
BREEAM (BRE Environmental Assessment Method)	Classification and best-of-class benchmarking	X	X	–	X	X	–	–	–	X	–	–	–	X	–	–	–	X	–	–	–
Building Innovation Fund	Contests, challenges and competitive grants	X	X	X	X	X	–	–	–	X	–	–	–	X	–	–	–	X	–	–	–
Capacity and Development Grant	Private regulation	X	X	X	X	–	–	–	X	–	–	–	X	X	–	–	–	–	X	–	–
Chicago Green Office Challenge	Contests, challenges and competitive grants	X	X	–	X	X	–	–	–	–	–	X	–	X	–	–	–	–	X	–	–
ClimateSmart Home Service	Intensive behavioural interventions	X	X	X	X	–	X	–	–	–	X	–	–	X	–	–	–	–	–	X	–

DGNB System (Deutsche Gütesiegel Nachhaltiges Bauen)	Classification and best-of-class benchmarking	X	X	–	X	X	–	–	–	X	–	–	–	X	–	–	–	X	–	–	–
Dutch Energy Service Company contracting	Innovative financing	X	X	X	X	X	–	–	–	–	X	–	–	X	–	–	–	X	–	–	–
E+Green Building	Contests, challenges and competitive grants	X	X	X	X	–	X	–	–	X	–	–	–	X	–	–	–	X	–	–	–
Eco-Office	Classification and best-of-class benchmarking	–	X	–	X	X	–	–	–	–	–	X	–	X	–	–	–	X	–	–	–
Energy Experience Programme	Intensive behavioural interventions	X	X	X	X	–	X	–	–	–	X	–	–	X	–	–	–	–	–	X	–
Environmental Upgrade Agreements	Innovative financing	X	X	X	X	X	–	–	–	–	X	–	–	X	–	–	–	X	–	–	–
ESCO Accreditation Scheme	Innovative financing	X	X	X	X	X	–	–	–	–	–	X	–	X	–	–	–	X	–	–	–
Grassroots Program	Contests, challenges and competitive grants	–	–	–	X	–	–	–	X	–	–	–	X	X	–	–	–	–	X	–	–
Green Building Index	Classification and best-of-class benchmarking	X	X	X	X	X	–	–	–	X	–	–	–	X	–	–	–	X	–	–	–
Green Labelling Scheme	Classification and best-of-class benchmarking	–	X	–	X	X	–	–	–	–	–	X	–	X	–	–	–	X	–	–	–
Green Leasing Toolkit	Green leasing	–	X	–	X	X	–	–	–	–	–	X	–	X	–	–	–	–	–	–	X

Name	Type	Role of government				Focus A				Focus B				Focus C				Focus D			
		G1	G2	G3	G4	A1	A2	A3	A4	B1	B2	B3	B4	C1	C2	C3	C4	D1	D2	D3	D4
Green Mark	Classification and best-of-class benchmarking	X	X	X	X	–	–	X	–	–	–	X	–	X	–	–	–	X	–	–	–
Green Star	Classification and best-of-class benchmarking	X	X	–	X	X	–	–	–	X	–	–	–	X	–	–	–	X	–	–	–
GRIHA (Green Rating for Integrated Habitat Assessment)	Classification and best-of-class benchmarking	X	X	X	X	X	–	–	–	X	–	–	–	X	–	–	–	X	–	–	–
International Green Construction Code	Private regulation	–	–	–	–	–	–	X	–	–	–	X	–	–	–	X	–	X	–	–	–
LEED (Leadership in Energy and Environmental Design)	Classification and best-of-class benchmarking	X	X	–	X	X	–	–	–	X	–	–	–	X	–	–	–	X	–	–	–
Local Law 86	Sustainable procurement	X	X	X	X	X	–	–	–	–	–	X	–	X	–	–	–	X	–	–	–
NABERS (National Australian Built Environment Rating System)	Classification and best-of-class benchmarking	X	X	X	X	X	–	–	–	X	–	–	–	X	–	–	–	X	–	–	–
National Green Leasing Policy	Green leasing	X	X	X	X	X	–	–	–	–	–	X	–	X	–	–	–	–	–	–	X
PACE (Property Assessed Clean Energy)	Innovative financing	X	X	X	X	–	–	X	–	–	–	X	–	X	–	–	–	X	–	–	–
Public Procurement Bill-2012	Sustainable procurement	X	X	X	X	X	–	–	–	–	–	X	–	X	–	–	–	X	–	–	–
Solar Leasing	Innovative financing	X	X	X	X	–	X	–	–	–	X	–	–	–	–	X	–	X	–	–	–

Sustainable Business Leader Program	Classification and best-of-class benchmarking	–	–	–	X	–	–	–	X	–	–	–	X	X	–	–	–	–	–	X	–
Sustainable Public Procurement	Sustainable procurement	X	X	X	X	X	–	–	–	–	–	X	–	X	–	–	–	X	–	–	–
Transition Towns	Private regulation	–	–	–	X	–	X	–	–	–	X	–	–	X	–	–	–	–	–	X	–

Note: Abbreviations used are G1 = government in an initiating or leading role; G2 = government in an assembling role; G3 = government in a guarding role. Focus A = what type of buildings does the tool address predominantly? A1 = commercial buildings; A2 = residential buildings; A3 = both commercial and residential buildings; A4 = other focus. Focus B = what type of buildings does the tool address predominantly? B1 = new buildings; B2 = existing buildings; B3 = both new and existing building; B4 = other focus. Focus C = what is the dominant focus of the tool? C1 = urban sustainability; C2 = urban resilience; C3 = both urban sustainability and resilience; C4 = other focus. Focus D = what is the dominant approach to seeking the tool's goal achievement? D1 = through technology; D2 = through behavioural change; D3 = through both technology and behavioural change; D4 = other focus.

Interview insights were complemented and contrasted with reported data on the various cases studied, such as policy reports, brochures and news articles (Golafshani, 2003; Mason, 2006; Seale et al., 2004). The interviews were highly valuable in that interviewees could provide me with documented information that was only available from their organization, for instance, policy briefs or annual reports. They further provided me with brochures, copies of news clippings and addresses to relevant web pages that I could not have obtained otherwise. Finally, through personal anecdotes and by drifting off from the semi-structured interview questions, they provided me with richer insights into the complexities of governing urban sustainability and resilience than I could have got from the questions I normally derive from the literature on this topic.

I have updated the datasets obtained from these earlier studies. I have predominantly done so by analysing the academic and other literature on the various cases that I discuss. I have further carried out extensive internet searches on the various cases reported on in the book. Through these I have been able to provide a richer insight into the various cases than in my earlier reported work.

That said, in this book I do not provide any direct quotes or give any direct references to any of the close to 500 interviews that underlie the earlier research projects. I have chosen to do so to maintain the anonymity that I promised my interviewees in these projects, and because the aim of this book is slightly different from the aim of these projects. Where the earlier research projects specifically sought to evaluate the performance of the cases, this book seeks to describe, map and evaluate the wide range of traditional and innovative governance tools for urban sustainability and resilience.

A.3 LIMITATIONS OF THE STUDY PRESENTED

The approach to this case selection and data gathering strategy has some limitations.

First, the selection of cases is limited to my earlier studies. That said, in this book I have followed up on and studied cases that have been suggested by participants in earlier research projects but that did not fully fit the aim of those projects. For instance, the various insurance-oriented tools discussed in this book have not appeared in my earlier work (for example, the Building Resilience Rating Tool, the National Flood Insurance Program). In building on my earlier work the sample of cases presented in this book is limited by the countries I have studied (see above).

I do not, therefore, claim that the sample of cases studied here is perfectly representative of the entire population of traditional and innovative governance tools for urban sustainability and resilience that is applied throughout the world. The size and diversity of this population simply is too large to make such a claim. I do, however, feel that the large set of countries and the large set of governance tools that I discuss in this book provide a window on the opportunities and constraints governments and NGOs face in governing urban sustainability and resilience. Even whilst it is not a perfectly representative sample of the range of the world's countries and governance tools (Hoffmann, 2011).

Second, in studying examples I have been limited by availability of data and my ability to trace examples. Most of these have been initially uncovered by internet searches, implying that governance tools that do not appear in such internet searches are not included in my sample, unless any of the participants in my studies referred to these.

In addition, my search for governance tools is limited by the languages that I master (Dutch, English and German). This, in particular, excludes a large range of local governance tools in non-English speaking countries in my study. Finally, the study is limited to the availability of interviewees. Even though I have been fortunate with the over 500 interviewees that had time to meet with me (and my collaborators in the study on energy performance of buildings in Europe), I have, for various reasons, not been able to meet with at least twice that number of individuals that I had targeted initially for interview in the various research projects.

Notes

CHAPTER 1

1. This book is interested in the governing of urban sustainability and resilience in cities, towns, villages and other urban environments. The words 'city' and 'urban environment' refer to these and are used interchangeably.
2. See World Bank (2008, 2009), the European Environmental Agency (EEA, 2012), the Indian Energy and Resources Institute (TERI, 2011), the International Council for Local Environmental Initiatives (ICLEI – Local Governments for Sustainability) (Otto-Zimmermann, 2010) and the government of New York City (City of New York, 2013).
3. In collaboration with Arcadis, the Netherlands.
4. Interviews with 103 individuals in Australia and Canada between 2007 and 2010 as part of a project I carried out on compliance with building codes; interviews with over 100 individuals between 2006 and 2011 in various projects I carried out on the reform of Dutch Building Codes; interviews with 108 individuals throughout the European Union in 2010 as part of a project on sustainable construction regulation in the 27 European Member States (I carried out this project in collaboration with Dr Ellen van Bueren and Arcadis, the Netherlands); and finally, interviews with 211 individuals in Australia, Germany, India, Malaysia, the Netherlands, Singapore, the United Kingdom and the United States between 2011 and 2013 as part of a project I carried out on urban sustainability.

CHAPTER 2

1. There are, of course, many more viewpoints of why governments are needed than the predominantly law and economics perspective provided here (Getimis, 2010; Supiot, 2007; Van Caenegem, 2003; Zeitlin, 1997).
2. The achievements of reduced greenhouse gas emissions in leading countries in, for instance, Europe can be critiqued because much polluting production has been exported to rapidly developing economies. How would Europe's greenhouse gas emissions look if not only locally produced emissions but also 'imported' emissions from consumption goods produced elsewhere were included in the equation (see Helm, 2012).
3. UN-HABITAT (2013) does not provide a direct definition of a slum but considers a slum area a settlement of households that cannot provide in one or more of the following living conditions: resilience against weather events; sufficient living space per individual; easy access to safe water; easy access to adequate sanitation; security of tenure that prevents forced evictions.

4. Throughout the book it will become clear that India provides an intriguing urban governance paradox that I hope to study in more depth in the years to come. Its government struggles with legalizing and regulating slums, while it is experimenting with truly innovative building regulation. Its property developers seek to overtake the United States in having the most square metres of sustainable built-up space voluntarily developed (see Chapter 4), while it has also developed the world's most expensive house (the Antilia building, estimated at US$1 billion) that overlooks many of Mumbai's slums.

CHAPTER 3

1. Collaborative governance is but one of the many streams of governance theorizing. It reflects one of the key aspects of governance that is found in many other streams of governance theorizing as well: the working together of governmental and non-governmental actors in addressing societal problems (for good and very accessible reviews of the governance literature, see Bell and Hindmoor, 2009; Chhotray and Stoker, 2010).

CHAPTER 4

1. Unfortunately, I could not gain more insight as to whether this 34 per cent energy reduction savings is coincidentally just below the 35 per cent split from where the ESCO needs to share these savings, or not.
2. As with all voluntary programmes, it is of course questionable how 'voluntary' this particular example is. Schools may voluntarily participate in the Energy Experience Programme, but once they choose to do so their students seem to have no choice but to participate.
3. Data from http://www.citymayors.com/development/built_environment_usa.html (accessed 26 February 2014).
4. Data from http://www.urbannewsdigest.in/green-cities/ (accessed 26 February 2014).
5. I appreciate that many high-performing buildings in the United States are not LEED (Platinum) certified, but even if I am wrong by a factor of 100 these numbers are still concerning, given that LEED is likely the world's most successful market-driven governance tool. All the more because so many of the other governance tools discussed in this book are in one way or the other linked to LEED.
6. Data from http://www.nytimes.com/ (accessed 26 February 2014).

CHAPTER 5

1. In distilling these design criteria for governance for urban sustainability and resilience I am strongly influenced by the work of Neil Gunningham, Peter Grabosky and Darren Sinclair, and in particular the conclusion to their book *Smart Regulation* (Gunningham and Grabosky, 1998).

2. In distilling these trends in contemporary governance for urban sustainability and resilience I am strongly influenced by Matthew Hoffmann's *Climate Change at the Crossroads* (Hoffmann, 2011).
3. I have actually introduced more examples than these 68, but to some I have given relatively limited attention. As the quantity of the number of examples discussed does not necessarily improve the quality of the insight that I seek to provide into governing urban sustainability and resilience, I feel that I cannot refer to some of the examples in a similar way as those introduced in Table 1.1.
4. Combined data from http://epp.eurostat.ec.europa.eu/ and http://en.wikipedia.org/wiki/List_of_cities_in_the_European_Union_with_more_than_100,000_inhabitants (accessed 10 February 2014).

APPENDIX

1. In collaboration with Dr Ellen van Bueren and Arcadis, the Netherlands.

References

Aakre, S. and J. Hovi (2010). 'Emission trading: participation enforcement determines the need for compliance enforcement'. *European Union Politics*, **11** (3), 427–45.

AASHE. (2013). Campus Sustainability Revolving Loan Funds Database, accessed 17 December 2013 at http://www.aashe.org/resources/campus-sustainability-revolving-loan-funds/.

Abaire, J. (2008). 'Green buildings: what it means to be "green" and the evolution of green building laws'. *The Urban Lawyer*, **40** (3), 623–32.

ABC News. (2012). *Solar Panel Subsidies Scrapped Early*, 16 November, accessed 13 November 2013 at http://www.abc.net.au/news/2012-11-16/solar-panels-subsidies-scrapped-early/4376520.

ABCB. (2004). *BCA 2004*. Canberra: Australian Building Codes Board.

ABCB. (2009). *Building Codes of Australia – Amdt 9*. Canberra: Australian Building Codes Board.

ABCB. (2011). *BCA 2011*. Canberra: Australian Building Codes Board.

ABCB. (2012). *Construction of Buildings in Flood Hazard Areas. Standard. Version 2012.2*. Canberra: Australian Building Codes Board.

Abramovitz, M. (1986). 'The privatization of the welfare state: a review'. *Social Work*, **31** (4), 257–64.

ACCC. (2011). *Continue to Beware of Scam Solar Offers*, September, accessed 13 November 2013 at http://www.scamwatch.gov.au/content/index.phtml/itemId/876741.

Ahern, J. (2013). 'Urban landscape sustainability and resilience: the promise and challenges of integrating ecology with urban planning and design'. *Landscape Ecology*, **28** (6), 1203–12.

Ahmed, S. and L. Lakhani (2013). 'Bangladesh building collapse: an end to recovery efforts, a promise of a new start', 14 June, accessed 11 February 2014 at http://edition.cnn.com/2013/05/14/world/asia/bangladesh-building-collapse-aftermath.

Albrito, P. (2012). 'Making cities resilient: increasing resilience to disasters at the local level'. *Journal of Business Continuity & Emergency Planning*, **5** (4), 291–7.

Aldinga Arts Eco Village. (2003). *Aldinga Arts Eco Village: By Laws*. Adelaide: Geoffrey Adam & Co.

Alexander, D. (1997). 'The study of natural disasters, 1977–97: some reflections on a changing field of knowledge'. *Disasters*, **21** (4), 284–304.

Alexander, J. (1987). 'The social requisites for altruism and voluntarism: some notes on what makes a sector independent. *Sociological Theory*, **5** (2), 165–71.

Alexander, L. (2010). 'Reflections on success and failure in new governance and the role of the lawyer'. *Wisconsin Law Review*, **2**, 737–48.

Allcott, H. (2011). 'Social norms and energy conservation'. *Journal of Public Economics*, **95** (9–10), 1082–95.

Alter, J. (2008). 'Slate on "decidedly dupable" LEED', accessed 2 January 20103 at http://www.treehugger.com/sustainable-product-design/slate-on-decidedly-dupable-leed.html.

Amecke, H. (2012). 'The impact of energy performance certificates: a survey of German home owners'. *Energy Policy*, **46** (1), 4–14.

Amir, O. and O. Lobel (2008). 'Stumble, predict, nudge: how behavioral economics informs law and policy'. *Columbia Law Review*, **108** (8), 2098–137.

Anand, C. and D. Aspul (2011). 'Economic and environmental analysis of standard, high efficiency, rainwater flushed, and composting toilets'. *Journal of Environmental Management*, **92** (3), 419–28.

Andrews, R.N.L. (1998). 'Environmental regulation and business self-regulation'. *Policy Science*, **31** (3), 177–97.

Ansell, C. and A. Gash (2008). 'Collaborative governance in theory and practice'. *Journal of Public Administration Research and Theory*, **18** (4), 543–71.

Arlington Economic Development. (n.d.). *Green Building Initiative*, accessed 14 November 2013 at http://www.arlingtonvirginiausa.com/major-initiatives/green-building-initiative/.

Armstrong, P. (2004). 'The Waste Wise Schools Program: evidence of educational, environmental, social and economic outcomes at the school and community level'. *Australian Journal of Environmental Education*, **20** (2), 1–11.

Arnstein, S.A. (1969). A ladder of citizen participation. *Journal of the American Institute of Planners*, **35** (4), 216–24.

Arora, S. and T.N. Cason (1995). 'An experiment in voluntary environmental regulation: participation in EPA's 33-50 program'. *Journal of Environmental Economics and Management*, **28** (3), 271–86.

Arora, S. and S. Gangopadhyay (1995). 'Toward a theoretical model of voluntary overcompliance'. *Journal of Economic Behavior and Organization*, **28** (3), 289–309.

Arup, RPA and Siemens. (2013). *Toolkit for Resilient Cities*. New York: Arup, RPA and Siemens.

Ash, M. and W. Ash (1899). *The Building Code of the City of New York as Constituted by the Greater New York Charter*. New York: Baker & Voorhis & Co.

Asimakopoulou, E. and N. Bessis (2010). *Advanced ICTs for Disaster Management and Threat Detection: Collaborative and Distributed Frameworks*. Hershey: IGI Global.

Atkinson, M. (2010). 'Buildings provide common ground for outcome in Cancun, in UNEP (ed.), *Climate Action: Assisting Business Towards Carbon Neutrality*, Nairobi: United Nations Environment Programme, pp. 32–7.

Australian Bureau of Statistics. (2010). *4613.0 – Australia's Environment: Issues and Trends*, 28 January, accessed 25 February 2014 at http://www.abs.gov.au/AUSSTATS/abs@.nsf/Lookup/4613.0Chapter75 Jan+2010.

Australian Bureau of Statistics. (2013). *4602.0.55.003 – Environmental Issues: Water Use and Conservation*, 30 October 2013, accessed 25 February 2014 at http://www.abs.gov.au/ausstats/abs@.nsf/media releasesbyReleaseDate/629A13C5A1CFAC3CCA2577DF00155272?Open Document.

Australian Government. (2008). *Energy Use in the Australian Residential Sector*. Canberra: Australian Government.

Australian Government. (2010). *Building Energy Efficiency Disclosure Act 2010*. Canberra: Commonwealth of Australia.

Australian Government. (2012). *Solar Credits for Small Generation Units*, accessed 13 November 2013 at http://www.climatechange.gov.au/reducing-carbon/renewable-energy/renewable-energy-target/small-scale-renewable-energy-systems/solar-credits-small-generation-units.

Australian Government. (2013a). *Home Energy Saver Scheme*, 7 November, accessed 18 December 2013 at http://www.dss.gov.au/our-responsibilities/communities-and-vulnerable-people/programs-services/financial-management-program/home-energy-saver-scheme-service-providers.

Australian Government. (2013b). *Sustainable Procurement Guide*. Canberra: Commonwealth of Australia.

Australian Resilience Taskforce. (2012). *Australian Resilience Taskforce: Overview and Functional Description*. Sydney: Australian Resilience Taskforce.

Australian Resilience Taskforce. (2013). *Creating a Less Brittle Built Environment*, accessed 11 February 2014 at http://www.building resilience.org.au/.

Ayres, I. and J. Braithwaite (1992). *Responsive Regulation. Transcending the Deregulation Debate*. New York: Oxford University Press.

Backstrand, K., J. Khan, A. Kronsell and E. Lovbrand (2010). *Environmental Politics and Deliberative Democracy: Examining the Promises of New Forms of Governance*. Cheltenham, UK and Northampton, MA, USA: Edward Elgar.

Baer, W.C. (1997). 'Towards design of regulations for the built environment'. *Environment and Planning B: Planning and Design*, **24** (1), 37–57.

Bailey, I. (2008). 'Industry environmental agreements and climate policy: learning by comparison'. *Journal of Environmental Policy & Planning*, **10** (2), 153–73.

Bakker, A. (ed.) (2009). *Tax and the Environment: A World of Possibilities*. Amsterdam: IBFD.

Balaras, C., A. Gaglia, E. Georgopoulou, S. Mirasgedis, Y. Srafidis and D. Lalas (2007). 'European residential buildings and empirical assessment of the Hellenic building stock, energy consumption, emissions and potential energy savings'. *Building and Environment*, **42** (3), 1298–314.

Baldwin, R. and J. Black (2008). 'Really responsive regulation'. *Modern Law Review*, **71** (1), 59–94.

Baldwin, R., B. Hutter and H. Rothstein (2000). *Risk Regulation, Management and Compliance*. London: London School of Economics.

Baldwin, R., M. Cave and M. Lodge (2011). *Understanding Regulation: Theory, Strategy and Practice*, 2nd edn. Oxford: Oxford University Press.

Bansal, P. and T. Hunter (2003). 'Strategic explanations for the early adoption of ISO 14001'. *Journal of Business Ethics*, **46** (3), 289–99.

Baranzini, A. and P. Thalmann (2004). *Voluntary Approaches in Climate Policy*, Cheltenham, UK and Northampton, MA, USA: Edward Elgar.

Barber, W.F. and R.V. Bartlet (2007). 'Problematic participants in deliberative democracy: experts, social movements, and environmental justice'. *International Journal of Public Administration*, **30** (1), 5–22.

Bardach, E. and R.A. Kagan (1982). *Going By the Book: The Problem of Regulatory Unreasonableness*. Philadelphia, PA: Temple University Press.

Baron, D.P. and D. Diermeier (2007). 'Strategic activism and nonmarket strategy'. *Journal of Economics and Management Strategy*, **16**, 599–634.

Barrett, S. (1991). 'Environmental regulation for competitive advantage'. *Business Strategy Review*, **2** (1), 1–15.

Bart, I.L. (2010). 'Urban sprawl and climate change: a statistical exploration of cause and effect, with policy options for the EU'. *Land Use Policy*, **27** (2), 283–92.

Bartle, I. and P. Vass (2007). 'Self-regulation within the regulatory state: towards a new regulatory paradigm?' *Public Administration*, **85** (4), 885–905.

Bartley, T. (2003). 'Certifying forests and factories: states, social movements, and the rise of private regulation in the apparel and forest products fields'. *Politics and Society*, **31** (3), 433–64. doi: 10.1177/0032329203254863.

BBC News. (2013). *Bangladesh Factory Collapse Toll Passes 1,000*, accessed 11 November 2013 at http://www.bbc.co.uk/news/world-asia-22476774.

BCA. (2008). *Code for Environmental Sustainability of Buildings*. Singapore: Building and Construction Authority

BCA. (2012). *BCA Green Mark. Certification Standards for New Buildings. Version 4.1*. Singapore: Building and Construction Authority.

BCA. (2013). *About BCA Green Mark Scheme*, accessed 12 December 2013 at http://www.bca.gov.sg/greenmark/green_mark_buildings.html.

Beatley, T. (2000). *Green Urbanism: Learning from European Cities*. Washington, DC: Island Press.

Beatley, T. (2009). *Green Urbanism Down Under*. Washington, DC: Island Press.

Bechberger, M. and D. Reiche (2004). 'Renewable energy policy in Germany: pioneering and exemplary regulations'. *Energy for Sustainable Development*, **8** (1), 47–57.

Beddoes, D. and C. Booth (2012). 'Insights and perceptions of sustainable design and construction' in C. Booth, F. Hammond, J. Lamond and D. Proverbs (eds), *Solutions for Climate Change Challenges in the Built Environment*. Oxford: Wiley-Blackwell, pp. 127–40.

Beerepoot, M. and N. Beerepoot (2007). 'Government regulation as an impetus for innovation: evidence from energy performance regulation in the Dutch residential building sector'. *Energy Policy*, **35**, 4812–25.

Beerepoot, M. and M. Sunikka (2005). 'The contribution of the EC energy certificate in improving sustainability of the housing stock'. *Environment and Planning B, Planning and Design*, **31** (1), 21–31.

Bell, S. and A. Hindmoor (2009). *Rethinking Governance*. Cambridge: Cambridge University Press.

Betsil, M. and H. Bulkeley (2006). 'Cities and the multilevel governance of global climate change'. *Global Governance*, **12** (2), 141–59.

Betsil, M. and H. Bulkeley (2007). 'Looking back and thinking ahead: a decade of cities and climate change research'. *Local Environment*, **12** (5), 447–56.

Bettencourt, L. and G. West (2010). 'A unified theory of uirban living'. *Nature*, **467** (7318), 912–13.

Better Buildings Partnership. (2013). *Better Buildings Partnership: Annual Report 2012–2013*. Sydney: City of Sydney.

Bhagavatula, L., C. Garzillo and R. Simpson (2013). 'Bridging the gap between science and practice: an ICLEI perspective'. *Journal of Cleaner Production*, **50** (3), 205–11.

Bhatty, A. (2010). 'Haiti devastation exposes shoddy construction', 15 January, accessed 11 February 2014 at http://news.bbc.co.uk/2/hi/8460042.stm.

Bird, S. and D. Hernandez, D. (2012). 'Policy options for the split incentive: increasing energy efficiency for low-income renters'. *Energy Policy*, **48**, 506–14.

BIS. (2005). *National Building Code of India 2005*. New Delhi: Bureau of Indian Standards.

Bischop, P. and G. Davis (2002). 'Mapping public participation in policy choices'. *Australian Journal of Public Administration*, **61** (1), 14–29.

Blackman, A., E. Uribe, B. Van Hoof and T.P. Lyon (2013). 'Voluntary environmental agreements in developing countries: the Colombian experience'. *Policy Sciences*, **46** (3), 335–85.

Blackstock, K. and C. Richards (2007). 'Evaluating stakeholder involvement in river basin planning: a Scottish case study'. *Water Policy*, **9** (5), 493–512.

Blinnikov, M., A. Shanin, N. Sobolev and L. Volkova (2006). 'Gated communities of the Moscow green belt: newly segregated landscapes and the suburban Russian environment'. *GeoJournal*, **66** (1–2), 65–81.

Blocken, B., W. Janssen and Y. Van Hooff (2012). 'CFD simulation for pedestrian wind comfort and wind safety in urban areas'. *Environmental Modelling & Software*, **30** (1), 15–34.

Bodansky, D., J. Brunnée and E. Hey (2008). *The Oxford Handbook of International Environmental Law*. Oxford: Oxford University Press.

Boonstra, W. (2013). 'Gemeente Amsterdam erkent falen investeringsfonds', *Binnenlands Bestuur*. 6 November.

Boorsma, B. and W. Wagener (2007). 'Connected urban development: innovation for sustainability'. *NATOA Journal*, 1–7.

Borghi, V. and R. Van Berkel (2007). 'New modes of governance in Italy and the Netherlands'. *Public Administration*, **85** (1), 83–101.

Bovens, M. (1998). *The Quest for Responsibility*. Cambridge: Cambridge University Press.

Boyd, S. (2013). 'Financing and managing energy projects through revolving loan funds'. *Sustainability*, **6** (6), 345–52.

Brabham, D. (2009). 'Crowdsourcing the public participation process for planning projects'. *Planning Theory*, **8** (3), 242–62.

Brady, H.E. and D. Collier (eds) (2004). *Rethinking Social Inquiry: Diverse Tools, Shared Standards*. Lanham, MD: Rowman & Littlefield.

Braithwaite, J. (2004). 'Methods of power for development: weapons of the weak, weapons of the strong'. *Michigan Journal of International Law*, **26** (1), 297–330.

Branker, K., M. Pathak and J. Pearce (2011). 'A review of solar photovoltaic levelized cost of electricity'. *Renewable and Sustainable Energy Reviews*, **15** (9), 4470–82.

BRE. (2013a). *BREEAM User Manual for the BREEAM In-Use Online System V2.0*. Watford: BRE Global.

BRE. (2013b). *Non-domestic – Government*, accessed 12 December 2013 at http://www.breeam.org/page.jsp?id=343.

BRE. (2013c). *Our History*, accessed 12 December 2013 at http://www.bre.co.uk/page.jsp?id=1712.

BREEAM. (2013). *BREEAM in Numbers*, accessed 10 December 2013 at http://www.breeam.org/page.jsp?id=559.

Breen, J. (1908). *De verordeningen op het bouwen te Amsterdam, voor de negentiende eeuw*. Amsterdam: Ten Brink and De Vries.

Bressers, H., T. De Bruijn and K. Lulofs (2009). 'Environmental negotiated agreement in the Netherlands'. *Environmental Politics*, **18** (1), 58–77.

Brevetti, F. (2008). 'Blue skies ahead for "gray water" systems?', *The Oakland Tribune*, 12 July.

Briomedia Green. (2012). *Singapore's Energy Services Companies (ESCOs) Accreditation Scheme*. Singapore: Briomedia Green.

Briscoe, F. and S. Safford (2008). 'The Nixon-in-China effect: activism, imitation, and the institutionalization of contentious practices'. *Administrative Science Quarterly*, **53** (3), 460–91.

Broeders, J. and J. Hakfoort (1999). '"And their right hand is full of bribes": corruption and real estate', in S.E. Roulac (ed.), *Ethics in Real Estate*. Boston, MA: Kluwer Academic Publishers, pp. 109–28.

Brooks, M. (2008). 'Green leases and green buildings'. *Probate & Property*, **14** (November/December), 23–6.

Brounen, D. and N. Kok (2011). 'On the economics of energy labels in the housing market'. *Journal of Environmental Economics and Management*, **62** (2), 166–79.

Brunei Times. (2009). *Indonesia's Quake Devastation Exposes Poor Building Standards*, accessed 11 February 2014 at http://www.bt.com.bn/news-asia/2009/10/09/indonesias-quake-devastation-exposes-poor-building-standards.

Butler, S. (2013). 'Retailers urged to take Bangladesh safety deal further', *Guardian*. 29 July, accessed 11 February 2014 at http://www.theguardian.com/world/2013/jul/28/retailers-bangladesh-safety-deal-further.

C40 Cities. (2013). *1200 Buildings Program*, accessed 16 December 2013 at http://www.c40cities.org/c40cities/melbourne/city_case_studies/1200-buildings-program.

Cadigan, J., P. Schmitt, R. Shupp and K. Swope (2011). 'The holdout problem and urban sprawl: experimental evidence'. *Journal of Urban Economics*, **59** (1), 72–81.

Cadman, D. (2007). *The Vicious Circle of Blame*. Bristol: University of the West of England.

Cafaggi, F. and A. Janczuk (2010). 'Private regulation and legal integration: the European example'. *Business and Politics*, **12** (3), 1–40.

California Department of HDC. (2009). *California Plumbing Code, Chapter 16*. Sacramento.

California Sustainability Alliance. (2009). *Greening California's Leased Office Space: Challenges and Opportunities*. San Francisco, CA: California Sustainability Alliance.

Campanella, T. (2006). 'Urban resilience and the recovery of New Orleans'. *Journal of the American Planning Association*, **72** (2), 141–6.

CAP. (2012). *Performance Audit on Jawaharlal Nehru National Urban Renewal Mission. Report No. 15 of 2012–13*. New Delhi: Comptroller and Auditor General of India.

Carbone, D. and J. Hanson (2013). 'Floods: 10 of the deadliest in Australian history', accessed 7 November 2013 at http://www.australiangeographic.com.au/journal/the-worst-floods-in-australian-history.htm.

Carr, C. (2013). 'PK Das on collaboratively remaking Mumbai', 22 July, accessed 29 November 2013 at http://www.urb.im/ca130722mme.

Casals, X. (2006). 'Analysis of building energy regulation and certification in Europe'. *Energy and Buildings*, **38** (3), 381–92.

Cashore, B., G. Auld and D. Newsom (2004). *Governing through Markets: Forest Certification and the Emergence of Non-state Authority*. New Haven, CT: Yale University Press.

CDFA. (2013). *Revolving Loan Funds*, accessed 17 December 2013 at http://www.cdfa.net/cdfa/cdfaweb.nsf/ordredirect.html?open&id=rlffactsheet.html.

Chakraborty, A. and A. Pfaelzer (2011). 'An overview of standby power management in electrical and electronic power devices and appliances to improve the overall energy efficiency in creating a green world'. *Journal of Renewable and Sustainable Energy*, **3** (2), 1–10.

Charles II. (1667). *An Act for Rebuilding the Citty of London. Statutes of the Realm. Volume 5: 1628–80*. London: British Parliament, reprinted in 1819.

CHBA. (2001). *Reform of Building Regulations – What do Members Think?* Ottawa: Canadian Home Builders Association.
Cheatham, C. (2009). 'New York City backs off retrofit requirement', accessed 17 November 2013 at http://www.greenbuildinglawupdate.com/2009/12/articles/codes-and-regulations/new-york-city-backs-off-retrofit-requirement/.
Chhotray, V. and G. Stoker (2010). *Governance Theory and Practice*. Houndmills: Palgrave.
Chicago Garden. (2012). *Chicago's Sustainable Backyard Program Workshop for Retailers*, 22 February, accessed 29 November 2013 at http://www.chicagonow.com/chicago-garden/2012/02/chicagos-sustainable-backyard-program-workshop-for-retailers/.
Chittock, D. and K. Hughey (2011). 'A review of international practice in the design of voluntary pollution prevention programs'. *Journal of Cleaner Production*, **19** (5), 542–51.
Chivell, W. (2005). *Inquest into the deaths of Johanna Paulina Maria Heynen and Marilyn Jeane McDougall*. Adelaide: Government of South Australia.
Chivers, J. and N. Flores (2002). 'Market failure in information: the National Flood Insurance Program'. *Land Economics*, **78** (4), 515–21.
Cialdini, R. (2009). *Influence: The Power of Persuasion*. New York: William Morrow.
Circo, C. (2008). 'Using mandates and incentives to promote sustainable construction and green building projects in the private sector: a call for more state land use policy initiatives'. *Penn State Law Review*, **112** (3), 731–44.
Cities Alliance. (2008). *Cities Alliance Annual Report 2008*. Brussels: Cities Alliance.
Cities Alliance. (2013). *Bombay First: Catalysing Urban Regeneration in Mumbai*, accessed 27 November 2013 at http://citiesalliance.org/node/1994.
City of Amsterdam. (2013). *Amsterdams Investeringsfonds*, accessed 17 December 2013 at http://www.amsterdam.nl/wonen-leefomgeving/klimaat-energie/amsterdams-investeri.
City of Boston. (2010). *Mayor Menino Announces 2010 Grassroots Program Funding Awards*, 22 September, accessed 18 December 2013 at http://www.cityofboston.gov/news/default.aspx?id=4774.
City of Boston. (2013a). *E+ Green Building Program*, accessed 17 December 2013 at http://www.epositiveboston.org/.
City of Boston. (2013b). *E+ Green Communities Program. Fact Sheet*. Boston: City of Boston.
City of Brisbane. (2006). *Brisbane CityShape 2026: The Draft*. Brisbane: Brisbane City Council.

City of Brisbane. (2009). *Growing a Green Heart Together*. Brisbane: City of Brisbane.

City of Brisbane. (2013a). *Brisbane's Draft New City Plan*, 6 November, accessed 29 November 2013 at http://www.brisbane.qld.gov.au/planning-building/planning-guidelines-and-tools/brisbanes-new-city-plan/index.htm.

City of Brisbane. (2013b). *The New City Plan: Community Consultation.* Brisbane: Brisbane City Council.

City of Melbourne. (2010). *1200 Buildings: Advice Sheet*. Melbourne: City of Melbourne.

City of Melbourne. (2013). *1200 Buildings: Current signatories*, accessed 16 December 2013 at http://www.melbourne.vic.gov.au/1200buildings/CurrentSignatories/Pages/CurrentSignatories.aspx.

City of New York. (2005). *Local Laws of the City of New York. No. 86.* New York: City of New York.

City of New York. (2009). *New York City Energy Conservation Code*. New York: City of New York.

City of New York. (2013). *PlanYC: A Stronger, More Resilient, New York.* New York: City of New York.

City of Rotterdam. (2011). *Factsheet Rotterdamse Groene Gebouwen – cluster zwembaden*. Rotterdam: City of Rotterdam.

City of Sydney. (2011). *Sustainable Sydney 2030*. Sydney: City of Sydney.

City of Sydney. (2013). *Smart Green Apartments: A Review of Initial Cost Savings and Community Improvement Resulting from the Smart Green Apartments Program*. Sydney: City of Sydney.

CitySwitch. (2013). *CitySwitch Green Office*, accessed 27 November 2013 at http://www.cityswitch.net.au/Home.aspx.

Clean Air Partnership. (2012). *Accelerating Adaptation in Canadian Communities*. Toronto: Clean Air Partnership.

Climate Group. (2008). *SMART 2020: Enabling the Low Carbon Economy in the Information Age*. London: The Climate Group on behalf of the Global eSustainability Initiative.

ClimateWise. (2012). *ClimateWise Principles: The Fifth Independent Review*. Cambridge: ClimateWise and University of Cambridge.

COAG. (2009). *National Strategy for Disaster Resilience: Building Our Nation's Resilience to Disasters*. Canberra: Council of Australian Governments.

Cobouw. (2003). 'Wijzigingen ontwerp bij bouw fataal', 12 June, *Cobouw*.

Cole, R. and M.J. Valdebenito (2013). The importation of building environmental certification systems: international usages of BREEAM and LEED. *Building Research & Information*, **41** (6), 662–76.

Collins, K. and R. Ison, R. (2009). 'Jumping off Arnstein's ladder: social learning as a new policy paradigm for climate change adaptation'. *Environmental Policy and Governance*, **19** (6), 358–73. doi: 10.1002/eet.523.

Communities and Local Government. (2006). *Code for Sustainable Homes: A Step-change in Sustainable Home Building Practice*. Wetherby: Communities and Local Government Publications.

Cook, L., H. Yan and S. Udas (2013). 'Mumbai mayor: decorator responsible for building collapse, killing 61', 30 September 2013, accessed 11 February 2014 at http://edition.cnn.com/2013/09/29/world/asia/mumbai-building-collapse/.

Cooper, I. and M. Symes (2009). *Sustainable Urban Development. Volume 4. Changing Professional Practice*. London: Routledge.

Corbett, C. and S. Muthulingam (2007). *Adoption of Voluntary Environmental Standards: The Role of Signaling and Intrinsic Benefits in the Diffusion of the LEED Green Building Standards*, unpublished manuscript.

Cork, S. (2010). *Resilience and Transformation*. Collingwood, Victora: CSIRO Publishing.

Croci, E. (2005). *The Handbook of Environmental Voluntary Agreements*. Dordrecht: Springer.

Croley, S. (2011). 'Beyond capture', in D. Levi-Faur (ed.), *Handbook on the Politics of Regulation*, Cheltenham, UK and Northampton, MA, USA: Edwar Elgar, pp. 50–69.

Darnall, N. and J. Carmin (2005). 'Greener and cleaner? The signaling accuracy of U.S. voluntary environmental programs', *Policy Sciences*, **38** (2–3), 71–90.

Darnall, N. and S. Sides (2008). 'Assessing the performance of voluntary environmental programs: does certification matter?' *Policy Studies Journal*, **36** (1), 95–117.

Das, P.K. (2012). *Open Mumbai: Re-envisioning the City and its Open Spaces*. Mumbai: P.K. Das & Associates.

Das, P.K. (2013). *Open Mumbai: Re-envisioning the City and its Open Spaces*, 18 August, accessed 29 November 2013 at http://www.thenatureofcities.com/2013/08/18/open-mumbai-re-envisioning-the-city-and-its-open-spaces/.

Davis, D. (2007). 'Should architects self certify building plans', 28 October, *Architectural Record*.

Davis, D. (2008). 'Debate ensues after NYC building chief resigns', 6 May, *Architectural Record*.

Davis, G. (2002). 'Policy capacity and the future of governance', in G. David and M. Keating (eds), *The Future of Governance*. St Leonards, New South Wales: Allen & Unwin, pp. 230–43.

Davis, M. (2011). *Behavior and Energy Savings: Evidence from a Series of Experimental Interventions*. New York: Environmental Defense Fund.

Dawe, G., R. Jucker and S. Martin (2005). *Sustainable Development in Higher Education: Current Practice and Future Developments*. York: The Higher Education Academy.

Dawson, S. (2004). 'Balancing self-interest and altruism: corporate governance alone is not enough'. *Corporate Governance*, **12** (2), 130–3.

De Almeida, A., P. Fonseca, B. Schlomann and N. Feilberg (2011). 'Characterization of the household electricity consumption in the EU, potential energy savings and specific policy recommendations'. *Energy and Buildings*, **43** (8), 1884–94.

De Bruijn, H., E. Ten Heuvelhof and M. Koopmans (2007). *Law Enforcement: The Game Between Inspectors and Inspectees*. Boca Raton, FL: Universal Publishers.

De Bruijn, T. and V. Norberg-Bohm (2005). *Industrial Transformation: Environmental Policy Innovation in the United States and Europe*. Cambridge, MA: MIT Press.

De Clercq, M. (2002). *Negotiating Environmental Agreements in Europe: Critical Factors for Success*. Cheltenham, UK and Northampton, MA, USA: Edward Elgar.

De Ranitz, J. (1948). *Het bouw- en woningtoezicht*. The Hague: VNG.

De Volkskrant. (2003). 'Balkons storten in: twee doden', 25 April, *de Volkskrant*, 1.

De Vreeze, N. (1993). *Woningbouw, inspiratie en ambities: Kwalitatieve grondslagen van de sociale woningbouw in Nederland*. Almere: Woningraad.

Dean, M. (2009). *Governmentality*. Los Angeles, CA: Sage.

Deflem, M. (2008). *Sociology of Law*. Cambridge: Cambridge University Press.

Delmas, M. A. and A. Keller (2005). 'Free riding in voluntary environmental programs: the case of the US EPA WasteWise Program'. *Policy Sciences*, **38** (2), 91–106.

Delmas, M.A. and A.K. Terlaak (2001). 'A framework for analyzing environmental voluntary agreements'. *California Management Review*, **43** (3), 44–62.

Delmas, M.A. and O.R. Young (2009). 'Introduction: new perspectives on governance for sustainable development', in M.A. Delmas and O.R. Young (eds), *Governance for the Environment: New Perspectives*. Cambridge: Cambridge University Press, pp. 3–11.

Deloitte. (2013). *Building Our Nation's Resilience to Natural Disasters*. Sydney: Deloitte Access Economics.

DeMarzo, P.M., M.J. Fishman and K.M. Hagerty (2005). 'Self-regulation and government oversight'. *The Review of Economic Studies*, **72** (3), 687–706.

Den Hond, F. and F.G.A. De Bakker (2007). 'Ideologically motivated activism: how activist groups influence corporate social change activities'. *Academy of Management Review*, **32** (3), 901–24.

Department of Standards Malaysia. (2007). *Malaysian Building Standards-Codes*. Putrajaya: Ministry of Science, Technology and Innovation.

Depietri, Y., F. Renaud and G. Kallis (2012). 'Heat waves and floods in urban areas: a policy-oriented review of ecosystem services'. *Sustainability Science*, **7** (1), 95–107.

Der Spiegel. (2008). 'Ich wollte alles hundertprozentig perfekt machen', *Der Spiegel*, 28 January.

Derissen, S., M. Quaas and S. Baumgartner (2011). 'The relationship between resilience and sustainability of ecological-economic systems'. *Ecological Economics*, **70** (6), 1121–8.

Derr, V., L. Chawla, M. Mintzer, D. Flanders Cushing and W. Van Vliet (2013). 'A city for all citizens: integrating children and youth from marginalized populations into city planning'. *Buildings*, **3** (4), 482–505.

DGBC. (2013). *Financing Tools for a Green Building Stock*. Rotterdam: Dutch Green Building Council.

DGNB. (2009). *German Sustainable Building Certificate – Second English Edition*. Stuttgart: German Sustainable Building Council.

DGNB. (2013). *International Application*, accessed 11 December 2013 at http://www.dgnb-system.de/en/system/international/.

Dietz, R., E. Ostrom and P. Stern (2003). 'The struggle to govern the commons'. *Science*, **302** (5652), 1907–12.

Dixon, L., N. Clancy, B. Bender, A. Kofner, D. Manheim and L. Zakaras (2013). *Flood Insurance in New York Following Hurricane Sandy*. Santa Monica, CA: RAND Corporation.

Dixon, T., M. Keeping and C. Roberts, C. (2008). 'Facing the future: energy performance certificates and commercial property'. *Journal of Property Investment & Finance*, **26** (1), 96–100.

Dixon, T., G. Ennis-Reynolds, C. Roberts and S. Sims (2009). 'Is there a demand for sustainable offices? An analysis of UK business occupier moves'. *Journal of Property Research*, **26** (1), 61–85.

Dobrev, S. and A. Gotsopoulos (2010). 'Legitimacy vacuum, structural imprinting, and the first mover disadvantage'. *Academy of Management Journal*, **53** (5), 1153–74.

Dodman, D. (2009). 'Blaming cities for climate change? An analysis of urban greenhouse gas emissions inventories'. *Environment and Urbanization*, **21** (1), 185–201.

Dohmen, F., M. Frohlingsdorf, A. Neubacher, T. Schulze and G. Traufetter, G. (2013). 'Germany's energy poverty: how electricity became a luxury good'. *SpiegelOnline*, 4 September.

Dolowitz, D.P. and D. Marsh (2000). 'Learning from abroad: the role of policy transfer in contemporary policy-making'. *Governance: An International Journal of Policy, Administration and Institutions*, **13** (1), 5–24.

Domask, J. (2003). 'From boycotts to global partnerships', in J.P. Doh and H. Teegen (eds), *Globalisation and NGOs*. Westport, CT: Praeger Publishers, pp. 157–86.

Domènech, L. and D. Saurí (2011). 'A comparative appraisal of the use of rainwater harvesting in single and multi-family buildings of the metropolitan area of Barcelona (Spain): social experience, drinking water savings and economic costs'. *Journal of Cleaner Production*, **19** (6–7), 598–608.

Drahos, P. (2004). 'Securing the future of intellectual property: intellectual property owners and their nodally coordinated enforcement pyramid'. *Case Western Reserve Journal of International Law*, **36** (1), 53–78.

Drahos, P. (2013). 'Rethinking the role of the patent office from the perspective of responsive regulation', in F. Abbott, C. Correa and P. Drahos (eds), *Emerging Markets and the World Patent Order*. Cheltenham, UK and Northampton, MA, USA: Edward Elgar, pp. 73–99.

Driessen, D. (2006). 'Economic instruments for sustainable development', in B. Richardson and S. Wood (eds), *Environmental Law for Sustainability*. Toronto: Hart Publishing, pp. 277–308.

Droege, P. (2008). *Urban Energy Transition: From Fossil Fuels to Renewable Power*. Oxford: Elsevier.

Dryzek, J. (2005). *The Politics of the Earth*, 2nd edn. Oxford: Oxford Universtity Press.

Duncan, J. (2005). 'Performance-based building: lessons from implementation in New Zealand'. *Building Research & Information*, **33** (2), 120–7.

Dunn, W.N. (2003). *Public Policy Analysis: An Introduction*. Harlow: Prentice Hall.

Dutch Government. (2013). *What is Sustainable Public Procurement?*, accessed 19 December 2013 at http://english.agentschapnl.nl/topics/sustainability/sustainable-procurement.

E2PO. (2013). *ESCO Accreditation Scheme*, 19 November, accessed 17 December 2013 at http://app.e2singapore.gov.sg/Programmes/ESCO_Accreditation_Scheme.aspx.

EC. (2000a). *Management of Construction and Demolition Waste*. Brussels: European Commission.

EC (2000b). Directive 2000/60/EC of the European Parliament and of the Council of 23 October 2000 establishing a framework for Community action in the field of water policy.

EC. (2006). *Directive 2006/32/EC (ESD)*. Brussels: European Parliament.

EC. (2008). *Directive 2008/98/EC (Waste Framework Directive)*. Brussels: European Parliament.

EC. (2010). *Directive 2010/31EU (EPBD – recast)*. Brussels: European Parliament.

EC. (2013). *Financial Support for Energy Efficiency in Buildings. Report SWD(2013) 143 final*. Brussels: European Commission.

Echenique, M., A. Hargreaves, G. Mitchell and A. Namdeo (2012). 'Growing cities sustainably: does urban form really matter?' *Journal of the American Planning Association*, **78** (2), 121–37.

EEA. (1997). *Environmental Agreements: Environmental Effectiveness*. Luxembourg: European Environmental Agency.

EEA. (2008). *Effectiveness of Environmental Taxes and Charges for Managing Sand, Gravel and Rock Extraction in Selected EU Countries*. Copenhagen: European Environment Agency.

EEA. (2012). *Challenges and Opportunities for Cities Together with Supportive National and European Policies*. Brussels: European Environment Agency.

EERI. (2009). *Learning from Earthquakes: The Mw 7.6 Western Sumatra Earthquake of September 30, 2009*. Oakland, CA: Earthquake Engineering Research Institute.

EIA. (2013). *How Much Electricity does an American Home Use?*, accessed 11 January 2014 at http://www.eia.gov/tools/faqs/faq.cfm?id=97&t=3.

Eichholtz, P., N. Kok and J. Quigley (2010). 'Doing well by doing good? Green office buildings'. *American Economic Review*, **100**, 2492–509.

Ekins, P. and E. Lees (2008). 'The impact of EU policies on energy use in and the evolution of the UK built environment'. *Energy Policy*, **36** (12), 4580–3.

Emden, A.C.R. (1885). *The Law Relating to Building, Building Leases, and Building Contracts with a Full Collection of Precedents with Respect to Matters Connected with Building, together with the Statute Law Relating to Building, with Notes and the Latest Cases under the Various Sections*. London: Stevens and Haynes.

Energy Matters. (2013). *Feed-in Tariff for Grid-connected Solar Power Systems*, accessed 13 November 2013 at http://www.energymatters.com.au/government-rebates/feedintariff.php.

Englehardt, J., T. Wu and G. Tchobaoglous (2013). 'Urban net-zero water treatment and mineralization: experiments, modeling and design'. *Water Research*, **74** (13), 4680–91.

Epstein, R. (1995). *Simple Rules for a Complex World*. Cambridge, MA: Harvard University Press.

European Commission. (2013). *Green Public Procurement*, accessed 19 December 2013 at http://ec.europa.eu/environment/gpp/versus_en.htm.

Evans, B., M. Joas, S. Sundback and K. Thobald (2005). *Governing Sustainable Cities*. London: Earthscan.

Evans, J., P. Jones and R. Krueger (2009). 'Organic regeneration and sustainability or can the credit crunch save our cities?' *Local Environment*, **14** (7), 683–98.

Evans-Cowely, J. and J. Hollander (2010). The new generation of public participation: internet-based participation tools. *Planning Practice & Research*, **25** (3), 397-408.

Eversole, R. (2010). 'Remaking participation'. *Community Development Journal*, **47** (1), 29–41.

Fairman, R. and C. Yapp (2005). 'Enforced self-regulation, prescription, and conceptions of compliance within small businesses: the impact of enforcement'. *Law and Policy*, **27** (4), 491–519.

Fay, R., G. Treloar and U. Iyer-Raniga (2000). 'Life-cycle energy analysis of buildings: a case study'. *Building Research & Information*, **28** (1), 31–41.

Feddersen, T.J. and T.J. Gilligan (2001). 'Saint and markets: activists and the supply of credence goods'. *Journal of Economics and Management Strategy*, **10** (1), 149–71.

Feynen, J., K. Shannon and M. Neville (2009). *Water & Urban Development Paradigms*. Leiden: CRC Press.

Flüeler, T. and H. Seiler (2003). 'Risk-based regulation of technical risks: lessons learnt from case studies in Switzerland'. *Journal of Risk Research*, **6** (3), 213–31.

Flynn, E. (2011). *Stanford University: The Building Energy Retrofit Programs. Green Revolving Funds in Action. Case Study Series*. Cambridge, MA: Sustainable Endowments Institute.

Foley, R. (2011). *Harvard University: Green Loan Fund. Green Revolving Funds in Action. Case Study Series*. Cambridge, MA: Sustainable Endowments Institute.

Folke, C., S. Carpenter, T. Elmqvist, L. Gunderson, C. Holling and B. Walker (2002). 'Resilience and sustainable development: building adaptive capacity in a world of transformations'. *AMBIO*, **31** (5), 437–40.

Ford, C. (2008). 'New governance, compliance, and principles-based securities regulation'. *American Business Law Journal*, 45 (1), 1–60.

Ford, C. and M. Condon (2011). New Governance and the Business Organization. *Law and Policy*, **33** (4), 449–58.

Forum for the Future. (2007). *Buying a Better World: Sustainable Public Procurement*. London: Forum for the Future.

Foucault, M. (2009). *The Birth of Biopolitics*. New York: Picador.

Fowler, K.M. and E.M. Rauch (2006a). *Sustainable Building Rating Systems*. Washington, DC: US Department of Energy.

Fowler, K.M. and E.M. Rauch (2006b). *Sustainable Building Rating Systems: Summary*. Richland, WA: Pacific Northwest National Laboratory.

Freeman, J. (1997). 'Collaborative governance in the administrative state'. *UCLA Law Review*, **45** (1), 1–98.

Freilich, R., R. Sitkowski and S. Mennillo (2010). *From Sprawl to Sustainablity: Smart Growth, New Urbanism, Green Development, and Renewable Energy*. Chicago, IL: American Bar Association.

Frondel, M., N. Ritter and C. Schmidt (2008). 'Germany's solar cell promotion: dark clouds on the horizon'. *Energy Policy*, **36** (11), 4198–204.

Frondel, M., N. Ritter, C. Schmidt and C. Vance, C. (2010). 'Economic impacts from the promotion of renewable energy technologies: the German experience'. *Energy Policy*, **38** (8), 4048–56.

Future Cities. (2013). *The Future Cities Guide: Accelerating Adaptation in Canadian Communities*. Essen: Future Cities.

Galiani, S. and E. Schargrodsky (2010). 'Property rights for the poor: effects of land titling'. *Journal of Public Economics*, **94** (9–10), 700–29.

Gandy, M. (2008). 'Landscapes of disaster: water, modernity, and urban fragmentation in Mumbai'. *Environment and Planning A*, **40** (1), 108–30.

Gann, D.M., Y. Wang and R. Hawkins (1998). 'Do regulations encourage innovation? The case of energy efficiency in housing'. *Building Research & Information*, **26** (4), 280–96.

Garvin, A. (2014). *The American City: What Works and What Doesn't*. New York: McGraw Hill.

GBCA. (2012). *A Decade in Green Building*. Sydney: Green Building Council of Australia.

GBCA. (2013a). *Valuing Green: How Green Buildings Affect Property Values and Getting the Valuation Method Right*. Sydney: Green Building Council of Australia.

GBCA. (2013b). *What is Green Star?*, 11 June, accessed 12 December 2013 at http://www.gbca.org.au/green-star/green-star-overview/what-is-green-star/2139.htm.

GBI. (2013). *Green Building Index*. Kuala Lumpur: Green Building Index SDN BHD.

GCF. (2009). *Global Construction 2020*. Oxford: Global Construction Perspectives and Oxford Economics.

Getimis, P. (2010). 'Strategic planning and urban governance: effectiveness and legitimacy', in M. Cerreta, G. Concillio and V. Monno (eds), *Making Strategies in Spatial Planning*. Dordrecht: Springer, pp. 123–46.

Ghadyalpatil, A. (2006). 'Officials are busy legalising slums', 29 September, *The Economic Times*, accessed 11 February 2014 at http://articles.economictimes.indiatimes.com/2006-09-29/news/27459167_1_ulhasnagar-regularising-slums.

Gifford, H. (2009). 'A better way to rate green buildings'. *Northeast Sun*, **27** (1), 19–27.

Gillingham, K., R. Newell and K. Palmer (2009). *Energy Efficiency Economics and Policy*. Cambridge, MA: National Bureau of Economics and Policy.

Glaeser, E. and M. Kahn (2010). 'The greenness of cities: carbon dioxide emissions and urban development'. *Journal of Urban Economics*, **67** (3), 404–18.

Glasbergen, P., F. Biermann and A. Mol (2007). *Partnerships, Governance and Sustainable Development*. Cheltenham, UK and Northampton, MA, USA: Edward Elgar.

Global Cities Covenant on Climate. (2013). *The Mexico City Pact. Second Annual Report 2012*. Mexico City: Global Cities Covenant on Climate.

Golafshani, N. (2003). 'Understanding reliability and validity in qualitative research'. *The Qualitative Report*, **8** (4), 597–607.

Goldman, C., N. Hopper and J. Osborn (2005). 'Review of US ESCO industry market trends: an empirical analysis of project data'. *Energy Policy*, **33**, 387–405.

Gould, E.R.L. (1895). *The Housing of the Working People*. Washington, DC: Government Printing Office.

Government of India. (2012). *The Public Procurement Bill 2012*. New Delhi: Commonwealth of India.

Government of Ontario. (2011). *Governments Work Together to Address Climate Change*, 7 January, accessed 25 November 2013 at http://

news.ontario.ca/ene/en/2011/01/governments-work-together-to-address-climate-change.html.

Government of SA. (2007). *Climate Change and Greenhouse Emissions Reduction Act 2007*. Adelaide: Government of South Australia.

Government of SA. (2009a). *Progress Report. Adelaide Brighton Cement Agreement*. Adelaide: Government of South Australia.

Government of SA. (2009b). *South Australian Commercial Property Sector Agreement*. Adelaide: Government of South Australia.

Government of SA. (2012). *Building Innovation Fund*, accessed 18 December 2013 at http://www.sa.gov.au/subject/Water,+energy+and+environment/Climate+change/Tackling+climate+change/What+ organisations,+business+and+industry+can+do/Building+Innovation+Fund.

Government Property Group. (2010). *National Green Leasing Policy*. Sydney: Ministrial Council on Energy.

Grabosky, P. (2013). 'Beyond responsive regulation: the expanding role of non-state actors in the regulatory process'. *Regulation & Governance*, **7** (1), 114–23.

Graff Zivin, J. and A. Small (2005). 'A Modigliani-Miller theory of altruistic corporate social responsibility'. *Topics in Economic Analysis and Policy*, **5** (1), 1–19.

Graham-Rowe, E., S. Skippon, B. Gardner and C. Abraham (2011). 'Can we reduce car use and, if so, how? A review of available evidence'. *Transportation Research Part A: Policy and Practice*, **45** (5), 401–18.

Gram-Hanssen, K., F. Bartiaux, O. Jensen and M. Cantaert (2007). 'Do homeowners use energy labels? A comparison between Denmark and Belgium'. *Energy Policy*, **35** (5), 2879–88.

Grant Thornton. (2011). *Appraisal of Jawaharlal Nehru National Urban Renewal Mission*. New Delhi: Grant Thornton India.

Green Billion. (2013). *The Billion Green Dollar Challenge*, accessed 17 December 2013 at http://greenbillion.org/.

Green Lease Library. (2013). *Green Lease Library*, accessed 12 December 2013 at http://www.greenleaselibrary.com/.

Green Mark. (2013). *About Green Mark*, accessed 12 December 2013 at http://www.greenmark.sg/about.html.

Green Strata. (2013). *Green Strata*, accessed 28 November 2013 at http://www.greenstrata.com.au/.

Greenhouse Office. (1999). *Australian Residential Building Sector Greenhouse Gas Emissions 1990–2010*. Canberra: Australian Greenhouse Office.

Greensense. (2013). *Out of Hours: The Easiest Way to Improve Your Building's Energy Efficiency*. Melbourne: Greensense.

GRIHA. (2014). *GRIHA Incentives*, accessed 25 February 2014 at http://www.grihaindia.org/index.php?option=com_content&view=article&id=109.

Gruvberger, C., H. Aspegren, B. Andersson and J. La Cour Jansen (2003). 'Sustainability concept for a newly built urban area in Malmö, Sweden'. *Water Science and Technology: A Journal of the International Association on Water Pollution Research*, **47** (7–8), 33–9.

Gsottbauer, E. and J. Van der Berg (2011). 'Environmental policy theory given bounded rationality and other-regarding preferences'. *Environmental and Resource Economics*, **49** (2), 263–304.

Guagnano, G. (2001). 'Altruism and market-like behavior: an analysis of willingness to pay for recycled paper products'. *Population and Environment*, **22** (4), 425–38.

Guerra Santin, O. and L. Itart (2012). 'The effect of energy performance regulations on energy consumption'. *Energy Efficiency*, **5** (3), 269–82.

Guest, D. (2011). Solar panel safety audit points the finger at rooftop cowboys, *The Australian*, 24 December, accessed 11 February 2014 at http://www.theaustralian.com.au/national-affairs/state-politics/solar-panel-safety-audit-points-the-finger-at-rooftop-cowboys/story-e6frgczx-1226229730324.

Gunawan, I. (2009). 'Earthquakes don't kill people, poorly constructed buildings do!', *The Jakarta Post*, accessed 11 February 2014 at http://www.thejakartapost.com/news/2009/12/03/earthquakes-don039t-kill-people-poorly-constructed-buildings-do.html.

Gunningham, N. (2009). 'The new collaborative governance: the localization of regulation'. *Journal of Law and Society*, **36** (1), 145–66.

Gunningham, N. and P. Grabosky (1998). *Smart Regulation: Designing Environmental Policy*. Oxford: Oxford University Press.

Gurjar, B.R., A. Jain, A. Sharma et al. (2010). 'Human health risks in megacities due to air pollution'. *Atmospheric Environment*, **44**, 4606–13.

Gurney, C. (1999). 'Pride and prejudice: discourses of normalisation in public and private accounts of home ownership'. *Housing Studies*, **14** (2), 163–83.

Gutmann, A. and D.F. Thompson (2004). *Why Deliberative Democracy?* Princeton, NJ: Princeton University Press.

Hamilton, J. (1995). 'Pollution as news: media and stock market reactions to the toxic inventory data'. *Journal of Environmental Economics and Management*, **28** (1), 98–113.

Hamin, E. and N. Gurran (2009). 'Urban form and climate change: balancing adaptation and mitigation in the U.S. and Australia'. *Habitat International*, **33** (3), 238–45.

Hansen, A.T. (1985). *The Regulation of Building Construction*. Ottawa: National Research Council Canada.

Harper, C. (2007). 'Climate change and tax policy'. *Boston College International & Comparative Law Review*, **30**, 411–60.

Hasofer, A., V. Beck and I. Bennets (2007). *Risk Analysis in Building and Fire Safety Engineering*. Oxford: Butterworth-Heinemann.

Hawkins, K. (1984). *Environment and Enforcement Regulation and the Social Definition of Pollution*. Oxford: Oxford University Press.

Heiskanen, E., M. Johnson, S. Robinson, E. Vadovics and M. Saastamoinen (2010). 'Low-carbon communities as a context for individual behavioural change'. *Energy Policy*, **38** (12), 7586–95.

Helm, D. (2012). *The Carbon Crunch*. New Haven, CT: Yale Universtity Press.

Hendriks, C. (2009). 'Deliberative governance in the context of power'. *Policy and Society*, **28** (3), 173–84.

Henryson, J., T. Hakansson and J. Pyrko (2000). 'Energy efficiency in buildings through information – Swedish perspective'. *Energy Policy*, **28** (3), 169–80.

Héritier, A. and S. Eckert (2008). 'New modes of governance in the shadow of hierarchy.' *Journal of Public Policy*, **28** (1), 113–38.

Hertier, A. (2002). *Common Goods: Reinventing European and International Governance*. Boston, MA: Rowman & Littlefield.

Hertier, A. and D. Lehmkuhl (2008). 'The shadow of hierarchy and new modes of governance'. *Journal of Public Policy*, **28** (1), 1–17.

Hettige, H., M. Huq, S. Pargal and D. Wheeler (1996). 'Determinants of pollution abatement in developing countries: evidence from South and Southeast Asia'. *World Development*, **24** (12), 1891–904.

Hickson, K. (2009). *The ABC of Carbon*. Toowong: ABC Carbon.

Hirokawa, K. (2009). 'At home with nature: early reflections on green building laws and the transformation of the built environment'. *Environmental Law*, **39** (2009), 507–77.

Hirschman, A.O. (1970). *Exit, Voice, and Loyalty: Responses to Decline in Firms, Organizations, and States*. Cambridge, MA: Harvard University Press.

HKGBC. (2013). *Certification*, accessed 11 December 2013 at http://www.hkgbc.org.hk/eng/certification.aspx.

HM Government. (2010). *The Building Regulations 2010: Part A*. Newcastle Upon Tyne: NBS.

Hochrainer, S. and R. Mechler (2011). 'Natural disaster risk in Asian megacities: a case for risk pooling?' *Cities*, **28** (1), 53–61.

Hodge, G.A. (2000). *Privatization: An International Review of Performance*. Boulder, CO: Westview Press.

Hoffman, A.J. (2001). *From Heresy to Dogma: An Institutional History of Corporate Environmentalism.* Stanford, CA: Stanford Business Books.

Hoffman, A.J. and R. Henn (2009). 'Overcoming the social barriers to green building.' *Organization & Environment*, **32** (4), 390–419.

Hoffmann, M. (2011). *Climate Governance at the Crossroads*. Oxford: Oxford University Press.

Hoffmann, W. (2006). 'PV solar electricity industry: market growth and perspective'. *Solar Energy Materials and Solar Cells*, **90** (18–19), 3285–311.

Hofman, H. (2013). 'Energieakkoord vraagt lef om van gebaande paden af te wijken', 22 February, accessed 17 December 2013 at http://www.energieakkoordser.nl/doen/blogs/blog-energieakkoord-vraagt-lef.aspx?p=1.

Hofman, P. and T. De Bruijn (2010). 'The emergence of sustainable innovation: key factors and regional support structures', in J. Sarking, J. Cordeiro and D. Vasquez Bruzt (eds), *Facilitating Sustainable Innovation through Collaboration*. Amsterdam: Springer, pp. 115–33.

Holley, C., N. Gunningham and C. Shearing (2012). *The New Environmental Governance*. London: Routledge.

Hong, W. and M.P. Laurenzi (2007). *Building Energy Efficiency: Why Green Buildings are Key to Asia's Future*. Hong Kong: Asia Business Council.

Hood, C., O. James, G. Jones, C. Scott and T. Travers (1998). 'Regulation inside government: where new public management meets the audit explosion'. *Public Money and Management*, **18** (2), 61–8.

Hood, C., H. Rothstein and R. Baldwin (2001). *The Government of Risk: Understanding Risk Regulation Regimes*. Oxford: Oxford University Press.

Horvat, M. and P. Fazio (2005). 'Comparative review of existing certification programs and performance assessment tools for residential buildings'. *Architectural Science Review*, **48** (1), 69–80.

Hossain, F. and J. Alam (2013). 'Bangladesh building collapse death toll tops 500; engineer whistleblower arrested', *Huffington Post*, 5 May, accessed 11 February 2014 at http://www.huffingtonpost.com/2013/05/02/bangladesh-death-toll-tops-500_n_3199568.html.

Howard, E. (1902). *Garden Cities of To-Morrow*. London: Sonnenschein & Co.

Howarth, R.B., B.M. Hadda and B. Paton (2000). 'The economics of energy efficiency: insights from voluntary participation programs'. *Energy Policy*, **28** (6–7), 477–86.

Howlett, M. (2011). *Designing Public Policies: Principles and Instruments*. Abingdon: Taylor and Francis.

Howlin, P., I. Magiati and T. Charman (2009). 'Systematic review of early intensive behavioral interventions for children with autism'. *American Journal on Intellectual and Developmental Disabilities*, **114** (1), 23–41.

Hunn, D. (2002). *Report of the Overview Group on the Weathertightness of Buildings to the Building Industry Authority. Submission of 31 August 2002*. Wellington: Building Industry Authority.

Hunold, C. (2001). 'Corporatism, pluralism, and democracy: toward a deliberative theory of bureaucratic accountability'. *Governance*, **14** (2), 151–67.

Hurst, D. (2012). 'Power-saving scheme axed', 26 April, accessed 11 February 2014 at http://www.brisbanetimes.com.au/queensland/power saving-scheme-axed-20120426-1xmyo.html.

ICC. (2006). *2006 International Building Code: Code & Commentary*. New York: Thomson Delmar Learning.

ICLEI. (2009). *Five Lessons from the Chicago Green Office Challenge*, accessed 17 December 2013 at http://www.icleiusa.org/blog/five-lessons-from-the-chicago-green-office-challenge.

IEA. (2001). *Things that Go Blip in the Night, Standby Power and How to Limit it*. Paris: International Energy Agency and OECD.

IEA. (2009). *World Energy Outlook 2009*. Paris: International Energy Agency and OECD.

IFRC. (2013). *World Disasters Report 2013*. Geneva: International Federation of Red Cross and Red Crescent Societies.

IGBC. (2013). *Indian Green Building Council*, accessed 25 December 2013 at http://www.igbc.in/site/igbc/index.jsp.

IIGCC. (2013). *Global Investor Survey on Climate Change: 3rd Annual Report on Actions and Progress*. London: Institutional Investors Group on Climate Change.

Imrie, R. (2004). 'The role of the building regulations in achieving housing quality'. *Environment and Planning B, Planning and Design*, **31** (3), 419–37.

Imrie, R. (2007). 'The interrelationships between building regulations and architects' practices'. *Environment and Planning B, Planning and Design*, **34** (5), 925–43.

Indvik, J., R. Foley and M. Orlowski (2013). *Green Revolving Funds: An Introductory Guide to Implementation & Management*. Cambridge, MA: Sustainable Endowments Institute.

International Code Council. (2010). *International Green Construction Code. Public Version 2.0*. Washington, DC: International Code Council.

International Code Council. (2012). 'First International Green Construction Code (IgCC) adoptions'. Fact Sheet. Washington, DC: International Code Council.

International Code Council. (2013). *International Green Construction Code*, accessed 13 December 2013 at http://www.iccsafe.org/cs/IGCC/Pages/default.aspx.

International Living Future Institute. (2014). *Understanding the Challenge*, accessed 10 February 2014 at http://living-future.org/node/117.

IPCC. (2014). *Climate Change 2014: Impacts, Adaptation, and Vulnerability.* Cambridge: Cambridge University Press.

Irvin, R. and J. Stansbury (2004). 'Citizen participation in decision making: is it worth the effort?' *Public Administration Review*, **64** (1), 55–65.

Irvine, H., K. Lazarevski and S. Dolnicar (2012). 'Strings attached: new public management, competitive grant funding and social capital'. *Financial Accountability & Management*, **25** (2), 225–52.

Ishikawa, J., K. Kiyono and M. Yomogida (2012). 'Is emission trading beneficial?' *Japanese Economic Review*, **63** (2), 185–203.

ISO. (2012a). *ISO & Construction*. Geneva: International Organization for Standardization.

ISO. (2012b). *Working Together – Partnerships for Positive Results*, 10 January, accessed 28 November 2013 at http://www.iso.org/iso/home/news_index/news_archive/news.htm?refid=Ref1521.

Jacobsson, S. and A. Bergek (2011). 'Innovation system analyses and sustainability transitions: contributions and suggestions for research'. *Environmental Innovation and Societal Transitions*, **1** (1), 41–57.

Jain, S. (2010). 'JNNURM: overview and appraisal'. *Context*, **7** (2), 4–8.

Janssen, R. (2005). *Towards Energy Efficient Buildings in Europe*. London: The European Alliance of Companies for Energy Efficiency in Buildings.

Jefferson, B., A. Laine, S. Parsons, T. Stephenson and S. Judd (2000). 'Technologies for domestic wastewater recycling'. *Urban Water*, **1** (4), 285–92.

Jelle, B.P., A. Hynd, A. Gustavsen, D. Arasteh and H. Goudey (2012). 'Fenestration of today and tomorrow: a state-of-the-art review and future research opportunities'. *Solar Energy Materials and Solar Cells*, **96** (1), 1–28.

Jenerette, G.D., S. Harlan, W. Stefanov and C. Martin (2011). 'Ecosystem services and urban heat riskscape moderation: water, green spaces, and social inequality in Phoenix, USA'. *Ecological Applications*, **21** (7), 2637–51.

Jo, J.H., J. Golden and H. Bryan (2010). 'Sustainable urban energy: development of a mesoscale assessment model for solar reflective roof technologies'. *Energy Policy*, **38** (12), 7951–9.

Jordan, B. (1989). *The Common Good: Citizenship, Morality, and Self-interest*. Oxford and New York: Blackwell.

Jordana, J. and D. Levi-Faur (2004). *The Politics of Regulation: Institutions and Regulatory Reforms for the Age of Governance*. Cheltenham, UK and Northampton, MA, USA: Edward Elgar.

Jurin, R. and L. Fox-Parrish (2008). 'Factors in helping educate about energy conservation'. *Applied Environmental Education & Communication*, **7** (3), 66–75.

Kagan, R.A. (1984). 'On regulatory inspectorates and police', in K. Hawkins and J.M. Thomas (eds), *Enforcing Regulation*. Boston, MA: Kluwer-Nijhoff, pp. 37–64.

Kaplow, S. (2009). 'Does a green building need a green lease?' *Baltimore Law Review*, **38** (3), 375–409.

Karathodorou, N., D. Graham and R. Noland (2010). 'Estimating the effect of urban density on fuel demand'. *Energy Economics*, **32** (1), 86–92.

Karkkainen, B. (2004). '"New governance" in legal thought and the world'. *Minnesota Law Review*, **89** (471–497), 77–99.

Kern, K. and G. Alber (2010). 'Governing climate change in cities', in L. Kamal-Chauoi (ed.), *Competitive Cities and Climate Change*. Milan: OECD, pp. 171–96.

Khanna, M. and W.R.Q. Anton (2002). 'Corporate environmental management: regulatory and market-based incentives'. *Land Economics*, **78** (4), 539–58.

Kibert, C.T. (2008). *Sustainable Construction: Green Building Design and Delivery*, 2nd edn, Hoboken, NJ: Wiley & Sons.

Kickert, W., E.-H. Klijn and J. Koppenjan (1997). *Managing Complex Networks*. London: Sage.

King, A.A. and M.J. Lenox (2000). 'Industry self-regulation without sanctions: the chemical industry's responsible care program'. *Academy of Management Journal*, **43** (4), 698–716.

King, B.G. (2008). 'A political mediation model of corporate response to social movement activism'. *Administrative Science Quarterly*, **53** (3), 395–421.

King, G., R.O. Keohane and S. Verba (1994). *Designing Social Inquiry: Scientific Inference in Qualitative Research*. Princeton, NJ: Princeton University Press.

King, L.W. (2004). *The Code of Hammurabi*. Whitfish, MT: Kessinger Publishing.

Kingsley, B. (2008). 'Making it easy to be green: using impact fees to encourage green building'. *New York University Law Review*, **83**, 532–45.

Kirkpatrick, C. and D. Parkers (2004). 'Regulatory impact assessment and regulatory governance in developing countries'. *Public Administration and Development*, **24** (4), 333–44.

Klinckenberg, F., M. Forbes Pirie and L. McAndrew (2013). *Renovation Roadmaps for Buildings*. London: The Policy Partners for Eurima.

Kocken, E.H.A. (2004). *Van bouwen, breken en branden in de lage landen; Oorsprong en ontwikkeling van het middeleeuws stedelijk bouwrecht tussen + 1200 en + 1500*. Deventer: Kluwer.

Kolokosta, D., D. Rovas, E. Kosmatopoulos and K. Kalaitzakis (2011). 'A roadmap towards intelligent net zero- and positive-energy buildings'. *Solar Energy Materials and Solar Cells*, **85** (12), 3067–84.

Konar, S. and M.A. Cohen (2001). 'Does the market value environmental performance?' *The Review of Economics and Statistics*, **83** (2), 281–9.

Konisky, D. and T. Beierle (2001). 'Innovations in public participation and environmental decision making'. *Society and Natural Resources*, **14** (9), 815–26.

KPMG. (2008). *A National Energy Program to Assist Low-income Households*. Sydney: KPMG.

Kwah, W.H., M. Ramli, K.J. Kam and M.Z. Sulieman (2012). 'Influence of the amount of recycled coarse aggregate in concrete design and durability properties'. *Construction and Building Materials*, **26** (1), 565–73.

Lankao, P.R. and H. Qin (2011). 'Conceptualizing urban vulnerability to global climate and environmental change'. *Current Opinion in Environmental Sustainability*, **3** (3), 142–9.

Larsen, P., C. Goldman and A. Satchwell (2010). 'Evolution of the U.S. energy service company industry: market size and project performance from 1990–2008'. *Energy Policy*, **50**, 802–20.

Lee, W.L. and J. Burnett (2008). 'Benchmarking energy use assessment of HK-BEAM, BREEAM and LEED'. *Building and Environment Building and Environment*, **43** (11), 1882–91.

Lee, W.L. and F.W.H. Yik (2004). 'Regulatory and voluntary approaches for enhancing building energy efficiency'. *Progress in Energy and Combustion Science*, **30** (5), 477–99.

Leichenko, R. (2011). 'Climate change and urban resilience'. *Current Opinion in Environmental Sustainability*, **3** (3), 164–8.

Lemke, T. (2002). 'Foucault, governmentality, and critique'. *Rethinking Marxism: A Journal of Economics, Culture & Society*, **14** (3), 49–64.

Lenox, M.J. and J. Nash. (2003). 'Industry self-regulation and adverse selection: a comparison accross four trade association programs'. *Business Strategy and the Environment*, **12** (6), 343–56.

Levi-Faur, D. (2011). *Handbook on the Politics of Regulation*. Cheltenham, UK and Northampton, MA, USA: Edward Elgar.

Lewis, J.M. (2011). 'The future of network governance research: strength in diversity and synthesis'. *Public Administration*, **89** (4), 1221–34.

LGA. (2013). *Summary Report. Climate Change Strategy Review*. Adelaide: Local Government Association of South Australia.

Lillie, N. and I. Greer (2007). 'Industrial relations, migration and neo-liberal politics: the case of the European construction sector'. *Politics and Society*, **35** (4), 551–81.

Lindsay, C.and R. McQuaid (2009). 'New governance and the case of activation policies'. *Social Policy & Administration*, **43** (5), 445–63.

LinkedIn. (2013). *Common Carbon Metric*, accessed 28 November 2013 at http://www.linkedin.com/groups/Common-Carbon-Metric-4836654/about.

Lipp, J. (2007). 'Lessons for effective renewable electricity policy from Denmark, Germany and the United Kingdom'. *Energy Policy*, **35** (11), 5481–95.

Listokin, D. and D.B. Hattis (2005). 'Building codes and housing'. *Cityscape: A Journal of Policy Development and Research*, **8** (1), 21–67.

Lloyd, G. (2013). 'Solar price rise to end power divide', 25 May, accessed 13 November 2013 at http://www.theaustralian.com.au/national-affairs/solar-price-rise-to-end-power-divide/story-fn59niix-1226650277855#.

Lobel, O. (2004). 'The renew deal: the fall of regulation and the rise of governance in contemporary legal thought'. *Minesota Law Review*, **89** (2), 263–93.

Longhurst, R. (2003). 'Semi-structured interviews and focus groups', in N.J. Clifford and G. Valentine (eds), *Key Methods in Geography*. London: Sage, pp. 117–32.

Lubin, D. and D. Esty (2010). 'The sustainability imperative'. *Harvard Business Review*, May, 2–8.

Lukensmeyer, C.J and L.H. Torres (2006). *Public Deliberation: A Manager's Guide to Citizen Engagement*. Washington, DC: IBM Center for the Business of Government.

Lynch, D. (2011). 'Turkish building collapses blamed on poor design code enforcement'. *The New Civil Engineer*, 1 November, accessed 11 February 2014 at http://www.nce.co.uk/news/international/turkish-building-collapses-blamed-on-poor-design-code-enforcement/8621913.article.

Lyon, T.P. (2009). 'Environmental governance: an economic perspective', in M.A. Delmas and O.R. Young (eds), *Governance for the Environment: New Perspectives*. Cambridge: Cambridge University Press, pp. 43–68.

Lyon, T.P. and J.W. Maxwell (2000). 'Voluntary approaches to environmental regulation: an overview', in M. Franzini and A. Nicita (eds), *Economic Institutions and Environmental Policy*. Aldershot: Ashgate, pp. 142–74.

Lyon, T.P. and J.W. Maxwell (2006). *Greenwash: Corporate Environmental Disclosure Under Threat of Audit*. Ann Arbor, MI: Ross School of Business.

Lyon, T.P. and J.W. Maxwell (2007). 'Environmental public voluntary programs reconsidered'. *Policy Studies Journal*, **35** (4), 723–50.

Ma, H., H. Shao and J. Song (2013). 'Modeling the relative roles of the foehn wind and urban expansion in the 2002 Beijing heat wave and possible mitigation by high reflective roofs'. *Meteorology and Atmospheric Physics*, October, 1–10.

MacLaren, V. (1996). 'Urban sustainability reporting'. *Journal of the American Planning Association*, **62** (2), 184–202.

Madigan, M. (2010). 'Shoddy solar power work poses risk of another subsidy debacle', 1 June, accessed 13 November 2013 at http://www.couriermail.com.au/business/shoddy-solar-power-work-poses-risk-of-another-subsidy-debacle/story-e6freqmx-1225874203000.

Magagna, V.V. (1988). 'Representing efficiency: corporate and democratic theory'. *The Review of Politics*, **50** (3), 420–44.

Maher, M. (2012). 'Local governance: Indian cities learn together, race ahead', *The Express Tribune*, 3 October, accessed 11 February 2014 at http://tribune.com.pk/story/446114/local-governance-indian-cities-learn-together-race-ahead/.

Mahoney, J. and G. Goertz (2006). 'A tale of two cultures: contrasting quantitative and qualitative research'. *Political Analysis*, **14**, 227–49.

Majone, G. (1990). *Deregulation or Re-regulation? Regulatory Reform in Europe and the United States*. London and New York: Pinter and St Martin's Press.

Majone, G. (1993). *Deregulation or Re-regulation? Policymaking in the European Community since the Single Act*. Florence: European University Institute.

Managan, K., J. Layke, A. Monica and C. Nesler (2012). *Driving Transformation to Energy Efficient Buildings*. Washington, DC: Institute for Building Efficiency.

Margaret. (2010). 'Chile's earthquake: an architect's perspective', 14 March, accessed 11 November 2013 at http://cachandochile.wordpress.com/2010/03/14/chiles-earthquake-an-arcitects-perspective/.

Martin, J. (2012). '3 kW solar PV systems: pricing, output, and returns', 16 July, accessed 13 November 2013 at http://www.solarchoice.net.au/blog/3kw-solar-pv-systems-pricing-output-and-returns/.

Maryland Energy Administration. (2012). *Maryland Green Building Tax Credit Program*, 7 May, accessed 14 November 2012 at http://energy.maryland.gov/Business/greenbuild.html.

Mason, J. (2006). 'Mixing methods in a qualitatively driven way'. *Qualitative Research*, **6** (1), 9–25. doi: 10.1177/1468794106058866.

Maxwell, J.W., T.P. Lyon and S.C. Hackett (2000). 'Self-regulation and social welfare: the political economy of corporate environmentalism'. *Journal of Law and Economics*, **43**, October, 583–618.

May, P. (1992). 'Policy learning or failure'. *Journal of Public Policy*, **12** (4), 331–54.

May, P. (2003). 'Performance-based regulation and regulatory regimes: the saga of leaky buildings'. *Law and Policy*, **25** (4), 381–401.

McAllister, I. and C. Sweett (2007). *Transforming Existing Buildings: The Green Challenge*. London: Royal Institute of Chartered Surveyors.

McGillian, C., P. De Wilde and S. Goodhew (2008). 'An assessment of the potential returns of energy certificates for the UK household sector'. *Journal of Financial Management of Property and Construction*, **13** (3), 187–99.

McGraw Hill. (2011). *Green Outlook 2011: Green Trends Driving Growth*. New York: McGraw Hill Construction.

McGraw Hill. (2012). *2013 Dodge Construction Green Outlook*. New York: McGraw Hill Construction.

McIntosh, A. and D. Wilkinson (2010). *The Australian Government's Solar PV Rebate Program: An Evaluation of its Cost-effectiveness and Fairness*. Canberra: The Australia Institute.

McKinsey. (2003). *Vision Mumbai: Transforming Mumbai into a World-class City*. Mumbai: McKinsey and Bombay First.

McLaughlin, K., S. Osborne and E. Ferlie (2002). *New Public Management: Current Trends and Future Prospects*. New York: Routledge.

McManus, P. (2005). *Vortex Cities to Sustainable Cities: Australia's Urban Challenge*. Sydney: UNSW Press.

Meacham, B., R. Bowen, J. Traw and A. Moore (2005). 'Performance-based building regulation: current situation and future needs'. *Building Research & Information*, **33** (2), 91–106.

Mikler, J. (2009). *Greening the Car Industry: Varieties of Capitalism in Climate Change*. Cheltenham, UK and Northampton, MA, USA: Edward Elgar.

Milman, A. and A. Short (2008). 'Incorporating resilience into sustainability indicators: an example for the urban water sector'. *Global Environmental Change*, **18** (4), 758–67.

Ministry of Economic Affairs. (2013). *Voortgangsrapportage Green Deals 2013*. The Hague: Ministerie van Economische Zaken.
Ministry of New and Renewable Energy. (2012). *The Little Book of GRIHA Rating*. New Delhi: Ministry of New and Renewable Energy.
Minnery, J., T. Argo, H. Winarso et al. (2013). 'Slum upgrading and urban governance: case studies in three South East Asian cities'. *Habitat International*, **39** (1), 162–9.
Mlenik, E., H. Visscher and A. Van Hal (2010). 'Barriers and opportunities for labels for highly energy-efficient houses'. *Energy Policy*, **38** (8), 4592–603.
Moe, E. (2012). 'Vested interests, energy efficiency and renewables in Japan'. *Energy Policy*, **40** (1), 260–73.
Mol, A., L. Volkmar and D. Liefferink (2000). *The Voluntary Approach to Environmental Policy: Joint Environmental Policy-making in Europe*. Oxford: Oxford University Press.
Moore, G. (2002). *Crossing the Chasm*. New York: HarperCollins.
Moore, S.A. (2007). *Alternative Routes to the Sustainable City*. Lanham, MD: Lexington Books.
Morgenstern, R. and W. Pizer (2007). *Reality Check: The Nature and Performance of Voluntary Environmental Programs in the United States, Europe and Japan*. Washington, DC: RFF Press.
Morrisey, J., T. Moore and R. Horne (2011). 'Affordable passive solar design in a temperate climate: an experiment in residential building orientation'. *Renewable Energy*, **36** (2), 568–77.
Mourada, K., J. Berndtsson and R. Berndtsson (2011). 'Potential fresh water saving using greywater in toilet flushing in Syria'. *Journal of Environmental Management*, **92** (10), 2447–53.
Murthy, N. (2010). *A Better India, a Better World*. New Delhi: Penguin India.
Myers, N. and J. Kent (2001). *Perverse Subsidies: How Tax Dollars can Undercut the Environment and the Economy*. Washington, DC: Island Press.
Mytelka, L. and K. Smith (2002). 'Policy learning and innovation theory: an interactive and co-evolving process'. *Research Policy*, **31** (8–9), 1467–79.
NABERS. (2013). *History*, accessed 12 December 2013 at http://www.nabers.gov.au/public/WebPages/ContentStandard.aspx?module=10&template= 3&include=History.htm&side=EventTertiary.htm.
Narayan, V. and B. Jain (2013). '3 BMC men held for Mazgaon crash'. *The Times of India*, 2 October, accessed 11 February 2014 at http://articles.timesofindia.indiatimes.com/2013-10-02/mumbai/42614544_1_building-collapse-illegal-alterations-bmc.

Nasar, J. (1999). *Design by Competition: Making Design Competition Work*. Cambridge: Cambridge University Press.

Nash, J. and R.L. Revesz (2007). 'Grandfathering and environmental regulation: the law and economics of new source review'. *Northwestern University Law Review*, **101** (4), 1677–734.

Nath, P. and B. Behera (2011). 'A critical review of impact of and adaptation to climate change in developed and developing economies'. *Environment, Development and Sustainability*, **13** (1), 141–62.

Natural Resources Canada. (2010). *Canadian Renewable and Conservation Expense (CRCE)*, 19 october, accessed 14 November 2013 at http://oee.nrcan.gc.ca/corporate/statistics/neud/dpa/policy_e/details.cfm?searchType=default%C2%A7oranditems=all|0max=10&pageId=1&categoryID=1%C2%AEionalDeliveryId=all&programTypes=4,5&keywords=&ID=974&attr=0.

Nayar, N. (2010). 'Bombay First. Public private partnership success story for regeneration of a megapolis', 26 April, accessed 27 November 2013 at http://www.citiesalliance.org/sites/citiesalliance.org/files/CA_Images/Bombay%20First%20Presentation%20for%2026%20April%202010_0.pdf.

NDTV. (2013). *Sharad Pawar Ticks Off Politicians, Says Pull Down Illegal Buildings in their Areas*, 13 April, accessed 11 November 2013 at http://www.ndtv.com/article/cities/sharad-pawar-ticks-off-politicians-says-pull-down-illegal-buildings-in-their-areas-353787.

NeJaime, D. (2009). 'When new governance fails'. *Ohio State Law Journal*, **70** (2), 323–99.

Netherlands Enerprise Agency. (2014). *Netherlands Enterprise Agency*, accessed 10 February 2014 at http://english.rvo.nl/.

New York Times. (2013). *$40 Million in Aid Set for Bangladesh Garment Workers, New York Times*, 23 December, accessed 26 February 2013 at http://www.nytimes.com/2013/12/24/business/international/40-million-in-aid-set-for-bangladesh-garment-workers.html?pagewanted=1.

Newman, P., T. Beatley and H. Boyer (2009). *Resilient Cities*. Washington, DC: Island Press.

Newsham, G., S. Mancini and B. Birt (2009). 'Do LEED-certified buildings save energy? Yes, but …' *Energy and Buildings*, **41** (8), 897–905.

Nichols, M. (2014, 31 January 2014). 'U.N. appoints former NYC Mayor Bloomberg cities, climate change envoy', accessed 11 February 2014 at http://www.reuters.com/article/2014/01/31/us-climate-un-bloomberg-idUSBREA0U02Q20140131.

Nijkamp, P. and H. Opschoor (1995). 'Urban environmental sustainability: critical issues and policy measures in a Third World context', in M.

Chatterji and Y. Kaizhong (eds), *Regional Science in Developing Countries*. New York: Macmillan, pp. 52–73.

Nill, J. and R. Kemp (2009). 'Evolutionary approaches for sustainable innovation policies: from niche to paradigm?' *Research Policy*, **38** (4), 668–80.

Nixon, P. (1976). 'The use of materials from demolition in construction'. *Resources Policy*, **2** (4), 276–83.

North Carolina General Assembly. (2008). *Senate Bill 1597/S.L. 2008-22*. Raleigh, NC: North Carolina General Assembly.

Noveck, B. (2011). 'The single point of failure', in S. Van der Hof and M. Groothuis (eds), *Innovating Government*. The Hague: Asserr Press, pp. 77–99.

Novotny, V. (2012). 'Water and energy link in cities of the future', in V. Lazarova, K.-H. Choo and P. Cornel (eds), *Water, Energy: Interaction of Water Use*. London: IWA Publishing, pp. 37–60.

NRC. (2003). 'In Maastricht moet je op je hoede zijn', 27 April, *NRC*, 3.

NSW Government. (2011). *National Australian Built Environment Rating System*. Sydney: New South Wales Government.

Nwabuzor, A. (2005). 'Corruption and development: new initiatives in economic openness and strengthened rule of law. *Journal of Business Ethics*, **59** (1–2), 121–38.

O'Flynn, J. and J. Wanna (eds) (2008). *Collaborative Governance: A New Era of Public Policy in Australia?* Canberra: ANU ePress.

OECD. (2003). *Voluntary Approaches for Environmental Policy*. Paris: OECD.

Office of Environment and Heritage. (2013). *Nabers Annual Report 2012–13*. Canberra: Office of Environment and Heritage.

Olson, M. (1965). *The Logic of Collective Action: Public Goods and the Theory of Groups*. Cambridge, MA: Harvard University Press.

OndernemendGroen. (2013). *Green Deals*, 20 November, accessed 27 November 2013 at http://www.ondernemendgroen.nl/GREENDEALS/Pages/default.aspx.

Osbaldiston, R. and J.P. Schott (2012). 'Environmental sustainability and behavioral science: meta-analysis of proenvironmental behavior experiments'. *Environment and Behavior*, **44** (2), 257–99.

Otterpohl, R. and C. Buzie (2011). 'Wastewater: reuse-oriented wastewater systems', in T. Letcher and D. Vallero (eds), *Waste: A Handbook for Management*. Burlington, MA: Academic Press, pp. 127–36.

Otto-Zimmermann, K. (2010). *Resilient Cities: Cities and Adaptation to Climate Change*. Dordrecht: Springer.

PACE Now. (2013a). *About PACE*, accessed 16 December 2013 at http://pacenow.org/about-pace/.

PACE Now. (2013b). *Annual Report*. Pleasantville, NY: PACE Now.

Parida, B., S. Iniyan and R. Goic (2011). 'A review of solar photovoltaic technologies'. *Renewable and Sustainable Energy Reviews*, **15** (3), 1625–36.

Parker, C. and V. Lehman Nielsen (2011). *Explaining Compliance: Business Responses to Regulation*. Cheltenham, UK and Northampton, MA, USA: Edward Elgar.

Parker, C., C. Scott, N. Lacey and J. Braithwaite (2005). *Regulating Law*. Oxford: Oxford University Press.

Parkin, M., M. Powell and K. Matthews (2005). *Economics*. Boston, MA: Addison-Wesley.

Peacock, F. (2013). *Solar Cowboys*, accessed 11 April 2013 at www.solarquaotes.com.au/blog/category/solar-cowboys/.

Pearce, D., G. Porter, R. Steenblik, J. Pieters and M. Potier (2003). *Envrionmentally Harmful Subsidies: Policy Issues and Challenges*. Paris: OECD.

PEARL. (2010). *Peer Experience and Reflective Learning*. New Delhi: National Institute of Urban Affairs.

PEARL. (2011). *Urban Initiatives. Volume 5*. New Delhi: National Institute of Urban Affairs.

PEARL. (2013). *Best Practices*, accessed 26 November 2013 at http://www.indiaurbanportal.in/BestPracticesResult.aspx?id=76&Type=4.

Pedro, J.B., F. Meijer and H. Visscher (2010). 'Building control systems of European Communion countries'. *International Joural of Law in the Built Environment*, **2** (1), 45–59.

Pennington, M., J. Gray, C. Donaldson, J. Walker and H. Dickinson (2011). 'Quantitative analysis at local and national level'. *Public Policy and Administration*, **27** (2), 145–67.

Perera, O., N. Chowdhury and A. Goswami (2007). *State of Play in Sustainable Public Procurement*. Winnipeg: International Institute for Sustainable Development and The Energy and Resources Institute.

Pérez-Lombard, L., J. Ortiz, R. González and I.R. Maestre (2009). 'A review of benchmarking, rating and labelling concepts within the framework of building energy certification schemes'. *Energy and Buildings*, **41** (3), 272–8.

Peters, K. (2010). 'Creating a sustainable urban agriculture revolution'. *Journal of Environmental Law and Litigation*, **25**, 203–47.

Pierson, P. (2004). *Politics in Time: History, Institutions, and Social Analysis*. Princeton, NJ: Princeton University Press.

Pinto, J., D. Cruz, A. Paiva, S. Pereira, P. Tavares and L. Fernandes (2012). 'Characterization of corn cob as a possible raw building material'. *Construction and Building Materials*, **34** (1), 28–33.

Pivo, G. (2010). 'Owner-tenant engagement in sustainable property investing'. *Journal of Sustainable Real Estate*, **2** (1), 183–99.

Pizer, W., R. Morgenstern and J.-S. Shih (2010). 'The performance of voluntary climate programs'. RFF Discussion Paper No. 08-13-REV. Resources for the Future, Washington, DC.

Pollak, J. and P. Slominski (2009). 'Experimentalist but not accountable governance?' *West European Politics*, **32** (5), 904–24.

Pope, C.A., M. Ezziati and D. Dockery (2009). 'Fine-particulate air pollution and life expectancy in the United States'. *New England Journal of Medicine*, **360** (4), 376–86.

Pope, J. and A. Owen (2009). 'Emission trading schemes: potential revenue effects, compliance costs and overall tax policy issues'. *Energy Policy*, **37** (11), 4595–603.

Popovic-Gerber, J., J.A. Oliver, N. Cordero et al. (2012). 'Power electronics enabling efficient energy usage: energy savings potential and technological challenges'. *IEEE Transactions on Power Electronics*, **27** (5), 2338–53.

Portney, K.E. (2003). *Taking Sustainable Cities Seriously*. Cambridge, MA: MIT Press.

Potoski, M. and A. Prakash (2005a). 'Covenants with weak swords: ISO 14001 and facilities environmental performance'. *Journal of Policy Analysis and Management*, **24** (4), 745–69.

Potoski, M. and A. Prakash (2005b). 'Green clubs and voluntary governance: ISO 14001 and firms' regulatory compliance'. *American Journal of Political Science*, **49** (2), 235–48.

Potoski, M. and A. Prakash (2009). *Voluntary Programs: A Club Theory Perspective*. Cambridge, MA: MIT Press.

Povoledo, E. (2009). 'Lax code enforcement seen in some building collapses in Italy', *New York Times*, 7 April, accessed 11 February 2014 at http://www.nytimes.com/2009/04/08/world/europe/08codes.html?_r=0.

Prakash, V. (2013). 'Death toll rises to 72 in Mumbai building collapse', 6 April, accessed 11 February 2014 at http://www.reuters.com/article/2013/04/06/us-india-mumbai-collapse-idUSBRE93502220130406.

Preuss, L. (2009). 'Addressing sustainable development through public procurement: the case of local government'. *Supply Chain Management*, **14** (3), 213–33.

Price, M.E. and S. Verhulst (2005). *Self-regulation and the Internet*. The Hague: Kluwer Law International.

Productivity Commission. (2012). *Barriers to Effective Climate Change Adaptation. Report No. 59*. Canberra: Commonwealth of Australia.

Purohit, K. (2013). 'Mumbai Mazgaon building collapse caused by poor quality construction material?', *The Hindustan Times*, 3 October, accessed 11 February 2014 at http://www.hindustantimes.com/india-news/mumbai/mumbai-mazgaon-building-collapse-caused-by-poor-quality-construction-material/article1-1130348.aspx.

Queensland Government. (2011). Green Door Information Paper. Queensland Government, Brisbane.

Queensland Reconstruction Authority. (2012). *Rebuilding a Stronger, More Resilient Queensland*. Brisbane: Queensland Government.

Rahman, A., J. Dbais and M. Imteaz (2010). 'Sustainability of rainwater harvesting systems in multistorey residential buildings'. *American Journal of Engineering and Applied Sciences*, **3** (1), 73–82.

Raman, S. and E. Shove (2000). 'The business of building regulation', in S. Fineman (ed.), *The Business of Greening*. Abingdon: Routledge, pp. 134–50.

Rawal, R., P. Vaidya, V. Ghatti et al. (2012). 'Energy code enforcement for beginners: a tiered approach to energy code in India'. Paper presented at the 2012 ACEEE Summer Study on Energy Efficiency in Buildings, Pacific Grove, acessed 11 February 2014 at http://www.aceee.org/files/proceedings/2012/data/papers/0193-000113.pdf.

Reams, M., K. Clinton and N. Lam (2012). 'Achievement of climate planning objectives among US member cities of the International Council for Local Environmental Initiatives (ICLEI)'. *Low Carbon Economy*, **3** (2), 137–43.

Rees, W. (2009). 'The ecological crisis and self-delusion: implications for the building sector'. *Building Research & Information*, **37** (3), 300–11.

Register, R. (1987). *Ecocity Berkeley: Building Cities for a Healthy Future*. Berkeley, CA: North Atlantic Books.

Reiche, D. (2005). *Handbook of Renewable Energies in the European Union*. Frankfurt am Main: Peter Lang.

Reid, E.M. and M.W. Toffel (2009). 'Responding to public and private politics: corporate disclosure of climate change strategies'. *Strategic Management Journal*, **30** (11), 1157–78.

Reiss, A. (1984). 'Selecting strategies of social control over organizational life', in K. Hawkins and J.M. Thomas (eds), *Enforcing Regulation*. Boston, MA: Kluwer-Nijhoff, pp. 23–35.

Reuters. (2013). *Nigeria Building Collapse Kills Four, Traps 50 Workers*, 4 November, accessed 11 November 2013 at http://www.reuters.com/article/2013/11/04/nigeria-collapse-idUSL5N0IP3R220131104.

Revell, A. and R. Blackburn (2007). 'The business case for sustainability? An examination of small firms in the UK's construction and restaurant sectors'. *Business Strategy and the Environment*, **16** (6), 404–20.

Rhodes, R.A.W. (1997). *Understanding Governance: Policy Networks, Governance, Reflexivity, and Accountability*. Buckingham and Philadelphia, PA: Open University Press.

Rhodes, R.A.W. (2007). 'Understanding governance: ten years on'. *Organization Studies*, **28** (8), 1243–64.

Richards, D. (1996). 'Elite interviewing: approaches and pitfalls'. *Politics*, **16** (3), 199–204.

Rihoux, B.and C. Ragin (2009). *Configurational Comparative Analysis*. London: Sage.

Risler, G. (1915). *Housing of the Working Classes in France. Cheap Up-to-date Dwellings in 1915. Exposition Universelle de San Francisco, 1915*. E´vreux: Impr. Ch. He´rissey.

Rivera, J. and P. de Leon (2004). 'Is greener whiter? Voluntary environmental performance of Western ski areas'. *Policy Studies Journal*, **32** (3), 417–37.

Robert, A. and M. Kummert (2012). 'Designing net-zero energy buildings for the future climate, not for the past'. *Building and Environment*, **55**, 150–8.

Robertson, I., H.R. Riggs, S. Yim and Y. Young (2007). 'Lessons from Hurricane Katrina storm surge on bridges and buildings'. *Journal of Waterway, Port, Coastal, and Ocean Engineering*, **133** (6), 463–83.

Robinson, W. and S. Min (2002). 'Is the first to market the first to fail? Empirical evidence for industrial goods businesses'. *Journal of Marketing Research*, **39** (1), 120–8.

Rockefeller Foundation. (2013). *100 Resilient Cities*, accessed 28 October 2013 at http://100resilientcities.rockefellerfoundation.org/.

Rogers, E. and E. Weber (2010). 'Thinking harder about outcomes for collaborative governance arrangements'. *American Review of Public Administration*, **40** (5), 546–67.

Romero, J. (2012). 'Blackouts illuminate India's power problems'. *IEEE Spectrum*, **49** (10), 11–12.

Romero-Lankao, P. and D. Dodman (2011). 'Cities in transition: transforming urban centers from hotbeds of GHG emissions and vulnerability to seedbeds of sustainability and resilience. Introduction and editorial overview'. *Current Opinion in Environmental Sustainability*, **13** (3), 113–20.

Roodman, D.M. and N. Lenssen (1995). *A Building Revolution: How Ecology and Health Concerns are Transforming Construction*. Washington, DC: Worldwatch Institute.

Rose, A. (2007). 'Economic resilience to natural and man-made disasters: multi-disciplinary origins and contextual dimensions'. *Environmental Hazards*, **7** (4), 383–98.

Rose, R. (2001). *Ten Steps in Learning Lessons from Abroad.* Hull: Future Governance Programme.

Roseland, M. (1997). 'Dimensions of the eco-city'. *Cities*, **14** (4), 197–202.

Rosenzweig, C., W. Solecki, S. Hammer and S. Mehrotra (2010). 'Cities lead the way in climate-change action'. *Nature*, **467** (7318), 909–11.

Rusk, B., T. Mahfouz and J. Jones (2011). 'Electricity's "disappearing act": understanding energy consumption and phantom loads'. *Technological Directions*, **71** (1), 22–5.

Sabatier, P.A. (2007). *Theories of the Policy Process.* Boulder, CO: Westview Press.

Sabel, C., A. Fung, B. Karkkainen, J. Cohen and J. Rogers (2000). *Beyond Backyard Environmentalism.* Boston, MA: Beacon Press.

Sabel, C. and W. Simon (2006). 'Accountability without sovereignity', in G. De Búrca and J. Scott (eds), *Law and New Governance in the EU and US.* Portland, OR: Hart Publishing, pp. 395–411.

Sadineni, S., S. Madala and R. Boehm (2011). 'Passive building energy savings: a review of building envelope components'. *Renewable and Sustainable Energy Reviews*, **15** (8), 3617–31.

Salop, S.C. and D.T. Scheffman (1991). 'Raising rivals' costs'. *American Economic Review*, **73** (2), 267–71.

Sanderson, I. (2002). 'Evaluation, policy learning and evidence-based policy making'. *Public Administration*, **80** (1), 1–22.

Sattar, T. (2013). 'Dhaka death trap fears follow fatal collapse', *Al Jazeera*, 5 May, accessed 11 February 2014 at http://www.aljazeera.com/indepth/features/2013/05/2013557235512347.html.

Saurwein, F. (2011). 'Regulatory choice for alternative modes of regulation: how context matters'. *Law and Policy*, **33** (3), 334–66.

Savan, B., C. Gore and A. Morgan (2004). 'Shifts in environmental governance in Canada: how are citizen environment groups to respond?' *Environment and Planning C*, **22** (4), 605–19.

SBLP. (2013). *Sustainable Business Leader Program*, accessed 11 December 2013 at http://sustainablebusinessleader.org/.

Scherba, A., D. Sailor, T. Rosenstiel and C. Wamser (2011). 'Modeling impacts of roof reflectivity, integrated photovoltaic panels and green roof systems on sensible heat flux into the urban environment'. *Building and Environment*, **46** (12), 2542–51.

Schindler, S.B. (2010). 'Following the industry's LEED: municipal adoption of private green building standards'. *Florida Law Review*, **62** (2), 285–350.

Schmeltz, M., S. González, L. Fuentes, A. Kwan, A. Ortega-Williams and L. Cowan (2013). 'Lessons from Hurricane Sandy: a community response in Brooklyn, New York'. *Journal of Urban Health*, **90** (5), 799–809.

Schmidt, T.M. and M. Fischlein (2010). 'Rival private governance networks: competing to define the rules of sustainability performance'. *Global Environmental Change*, **20** (3), 511–22.

Schout, A., A. Jordan and M. Twena (2010). 'From "old" to "new" governance in the EU'. *West European Politics*, **33** (1), 154–70.

Scofield, J. (2009). 'Do LEED-certified buildings save energy? Not really…' *Energy and Buildings*, **41** (12), 1386–90.

Scofield, J. (2013). 'Efficacy of LEED-certification in reducing energy consumption and greenhouse gas emissions for large New York City office buildings'. *Energy and Buildings*, **67**, 517–24.

Scott, C. (2009). 'Governing without law or governing without government?' *European Law Journal*, **15** (2), 160–73.

Scott, J., G. Adams and B. Wechsler (2004). 'Deliberative governance', in P. Bogason, S. Kensen and H. Miller (eds), *Tampering with Tradition*. Lanham, MD: Lexington Books, pp. 11–22.

Seale, C., G. Gobo, J.F. Gubrium and D. Silverman(2004). *Qualitative Research Practice*. London: Sage.

Segerson, K. and T. Miceli (1998). 'Voluntary environmental agreements: good or bad news for environmental protection?' *Journal of Environmental Economics and Management*, **36** (2), 109–30.

Sen, J. (2013). *Sustainable Urban Planning*. New Delhi: The Energy and Resource Institute.

Senate Standing Committees on Environment and Communications. (2013). *Recent Trends in and Preparedness for Extreme Weather Events*. Canberra: Commonwealth of Australia.

Sengers, F., R. Raven and A. Van Venrooij (2010). 'From riches to rags: biofuels, media discourses, and resistance to sustainable energy technologies'. *Energy Policy*, **38** (9), 5013–27.

Seville, C. (2011). 'How to cheat at LEED for homes', 24 May, accessed 10 December 2013 at http://www.greenbuildingadvisor.com/blogs/dept/green-building-curmudgeon/how-cheat-leed-homes.

Seyad, A., S. Baeke and M. De Clercq (1998). 'Success determining fators for negotiated agreements', in P. Glasbergen (ed.), *Co-operative Environmental Governance.* Amsterdam: Kluwer, pp. 111–32.

Seyfang, G. (2009). *Green Shoots of Sustainability*. Norwich: University of East Anglia.

Shapiro, S. (2009). 'Code green: is "greening" the building code the best approach to create a sustainable built environment?' *Planning & Environmental Law*, **63** (6), 3–12.

Shavell, S. (2007). 'On optimal legal change, past behaviour, and grandfathering'. *NBER Working Paper Series,* Working Paper No. 13563, 1–36.

Shimkada, R., J. Peabody, S. Quimbo and O. Solon (2008). 'The Quality Improvement Demonstration Study'. *Health Research Policy and Systems*, **6** (5), 1–12.

Short, J. and M.W. Toffel (2010). 'Making self-regulation more than merely symbolic: the critical role of the legal environment. *Administrative Science Quarterly*, **55** (2), 361–96.

Sichtermann, J. (2011). 'Slowing the pace of recovery: why property assessed clean energy programs risk repeating the mistakes of the recent foreclosure crisis'. *Valparadiso University Law Review*, **46**, 263–309.

Siddiquee, N.A. (2010). 'Combating corruption and managing integrity in Malaysia: a critical overview of recent strategies and initiatives'. *Public Organization Review*, **10** (2), 153–71.

Silverman, D. (2001). *Interpreting Qualitative Data*, 2nd edn. London: Sage.

Simon, H.A. (1945). *Administrative Behavior: A Study of Decision-making Processes in Administrative Organization.* New York: Free Press.

Simons, W. (2013). '6 vragen aan ESCo-specialist Albert Hulshoff', 29 August, accessed 17 December 2013 at http://www.energievastgoed.nl/2013/08/6-vragen-aan-esco-specialist-albert-hulshoff-je-wordt-als-gebouweigenaar-met-een-esco-volledig-ontzorgd/.

Sinclair, D. (1997). 'Self-regulation versus command and control? Beyond false dichotomies'. *Law and Policy*, **19** (4), 529–59.

Singapore Environment Council. (2013a). *Project ECO-Office*, accessed 11 December 2013 at http://www.ecooffice.com.sg/web/index.php.

Singapore Environment Council. (2013b). *The Singapore Green Labelling Scheme*, accessed 11 February 2014 at http://www.sec.org.sg/sgls/.

Singapore Power. (2013). *Energy Experience Programme*, accessed 18 December 2013 at http://www.singaporepower.com.sg/irj/servlet/prt/portal/prtroot/docs/guid/b080b6ca-1a7d-2e10-daa1-dba2df443292?spstab=Energy%20Efficiency.

Singleton, S. (2002). 'Collaborative environmental planning in the West: the good, the bad and the ugly'. *Environmental Politics*, **11** (3), 54–75.

Skocpol, T. (1985). 'Bringing the state back in: strategies of analysis in current research', in P. Evans, D. Rueschemeyer and T. Skocpol (eds), *Bringing the State Back In*. Cambridge: Cambridge University Press, pp. 3–37.

SMART 2020. (2013). *Connected Urban Development*, accessed 28 November 2013 at http://www.smart2020.org/case-studies/connected-urban-development/.

Smismans, S. (2008). 'New modes of governance and the participartory myth'. *West European Politics*, **31** (5), 874–95.

Smith, A. (2011). 'Cowboy traders blighting solar power industry', *The Sydney Morning Herald*, 15 October, accessed 11 February 2014 at http://www.smh.com.au/environment/energy-smart/cowboy-traders-blighting-solar-power-industry-20111014-1lp7y.html.

Smith, A., A. Stirling and F. Berkhout (2005). 'The governance of sustainable socio-technical transitions'. *Research Policy*, **34** (10), 1491–510.

Smith, K. (2013). *Environmental Hazards: Assessing Risk and Reducing Disaster*. New York: Routledge.

Smith, T. and M. Fischlein (2010). 'Rivalry private governance networks: competing to define the rules of sustainability performance'. *Global Environmental Change*, **20** (3), 511–22.

Sohail, M. and S. Cavill (2008). 'Accountability to prevent corruption in construction projects'. *Journal of Construction Engineering and Management*, **134** (9), 729–38.

Solanki, P.S., V.S. Malella and C. Zhou (2013). 'An investigation of standby energy losses in residential sector: solutions and policies'. *International Journal of Energy and Environment*, **4** (1), 117–26.

Solomon, J. (2008). 'Law and governance in the 21st century regulatory state'. *Texas Law Review*, **86**, 819–56.

Sorrell, S., E. O'Malley, J. Schleich and S. Scott (2004). *The Economics of Energy Efficiency*. Cheltenham, UK and Northampton, MA, USA: Edward Elgar.

Sparrow, M.K. (2000). *The Regulatory Craft: Controlling Risks, Solving Problems, and Managing Compliance*. Washington, DC: Brookings Institution.

Specht, K., R. Siebert, I. Hartmann et al. (2013). 'Urban agriculture of the future: an overview of sustainability aspects of food production in and on buildings'. *Agriculture and Human Values*, **30**, May, 1–19.

Spence, R. (2004). 'Risk and regulation: can improved government action reduce the impacts of natural disasters?' *Building Research & Information*, **32** (5), 391–402.

SPRING. (2012). *Grow Your Business through the Capability Development Grant*. Singapore: SPRING Singapore.

Starossek, U. (2006). 'Progressive collapse of structures: nomenclature and procedures'. *Structural Engineering International*, **16** (2), 113–17.

Stern, N. (2006). *Stern Review on the Economics of Climate Change*. London: HM Treasury.

Steurer, R. (2010). 'The role of governments in corporate social responsibility'. *Policy Science*, **43** (1), 49–72.
Steurer, R. (2013). 'Disentangling governance: a synoptic view of regulation by government, business and civil society'. *Policy Science*, **46** (4), 387–410.
Stewart, R. (2006). 'Instrument choice', in D. Bodansky, J. Brunnée and E. Hey (eds), *The Oxford Handbook of Environmental Law.* Oxford: Oxford University Press, pp. 147–81.
Stirling, A. (2004). 'Opening up or closing down? Analysis, participation and power in the social appraisal of technology', in M. Leach, I. Scoones and B. Wynne (eds), *Science and Citizens: Globalisation and the Challenge of Engagement.* New Delhi: Orient Longman, pp. 218–31.
Stren, R., R. White and J. Whitney (1992). *Sustainable Cities: Urbanization and the Environment in International Perspective.* Boulder, CO: Westview Press.
Sunstein, C.R and R. Hastie (2008). *Four Failures of Deliberating Groups.* Chicago, IL: Law School, University of Chicago.
Supiot, A. (2007). *Homo juridicus on the Anthropological Function of the Law.* London: Verso.
Sustainable Endowment Institute. (2012). *Greening the Bottom Line.* Cambridge, MA: Sustainable Endowment Insitute.
Swiss RE. (2013). *Mind the Risk.* Zurich: Swiss RE.
Taber, F. and N. Taylor (2009). 'Climate of concern – a search for effective strategies for teaching children about global warming'. *International Journal of Environmental and Science Education*, **4** (2), 97–116.
Takewaki, I., K. Fujita, K. Yamamoto and H. Takabatake (2011). 'Smart passive damper control for greater building earthquake resilience in sustainable cities'. *Sustainable Cities and Society*, **1** (1), 3–15.
Teisman, G.R. and E.-H. Klijn (2002). 'Partnership arrangements: governmental rhetoric or governance scheme? *Public Administration Review*, **62** (2), 197–205.
Ten Brink, P. (2002). *Voluntary Environmental Agreements: Process, Practice and Future Use.* Sheffield: Greenleaf.
Ten Hoeve, J. and M. Jacobson (2012). 'Worldwide health effects of the Fukushima Daiichi nuclear accident'. *Energy and Environmental Science*, **5**, 8743–57.
TERI. (2008). *Sustainable Public Procurement.* Winnipeg: International Institute for Sustainable Development.
TERI. (2011). *Mainstreaming Urban Resilience Planning in Indian Cities – A Policy Perspective.* New Delhi: The Energy and Resources Institute.

Termeer, C. (2009). 'Barriers to new modes of horizontal governance'. *Public Management Review*, **11** (3), 299–316.

The Economist. (2013). *Disaster at Rana Plaza*, 4 March, accessed 11 February 2014 at http://www.economist.com/news/leaders/21577067-gruesome-accident-should-make-all-bosses-think-harder-about-what-behaving-responsibly.

The Hindu. (2005). *Slums till 2000 to be Legalised in Mumbai*, *The Hindu*, 9 March, at http://www.hindu.com/2005/03/09/stories/2005030906500300.htm.

The Japan Times. (2005). 'Who checks the checkers? System allowing architect to pick lax inspectors of building plans'. *The Japan Times*, 25 November.

The Yomiuri Shimbun. (2005). 'Contractors to demolish 13 suspect housing blocks', *The Yomiuri Shimbun*, 24 November.

Thirani, N. (2012). 'In Mumbai, open spaces are rare, and rarely open'. *International New York Times*, 3 September, accessed 11 February 2014 at http://india.blogs.nytimes.com/2012/09/03/in-mumbai-open-spaces-are-rare-and-rarely-open/?_r=0.

Thomas, S. and N. Yeshwantrao (2013). 'Govt knew Thane district had 5 lakh illegal buildings, but did little'. *The Times of India*, 6 April, accessed 11 February 2014 at http://articles.timesofindia.indiatimes.com/2013-04-06/mumbai/38326763_1_illegal-structures-illegal-construction-construction-cost.

Thompson, D.F. (1980). 'Moral responsibility of public officials: the problem of many hands'. *American Political Science Review*, **74** (4), 905–16.

Timmer, C.P. (2012). 'Behavioral dimensions of food security'. *Proceedings of the National Academy of Sciences of the USA*, **109** (31), 12315–20.

Tommel, I. and A. Verdun (2009). *Innovative Governance in the European Union: The Politics of Multilevel Policymaking*. Boulder, CO: Lynne Rienner.

Transition Network. (2013a). *About Us*, accessed 13 December 2013 at http://www.transitionnetwork.org/.

Transition Network. (2013b). *Resources*, accessed 13 Decmber 2013 at http://www.transitionnetwork.org/resources?page=1.

Transparency International. (2009). *Overview of Corruption and Anti-corruption Efforts in India*. Berlin: Transparency International.

Trubek, D. and L.G. Trubek (2007). 'New governance & legal regulation: complementary, rivalry, and transformation'. *Columbia Journal of European Law*, **13** (3), 539–64.

Tweede Kamer der Staten Generaal. (2012). *Kamerstuk 33124*. The Hague: Sdu Uitgevers.

Tyler, T.R. (1990). *Why People Obey the Law*. New Haven, CT: Yale University Press.

Udas, S. (2013). 'Death toll rises to 72 in India building collapse', accessed 11 February 2014 at http://edition.cnn.com/2013/04/06/world/asia/india-fatal-bulding-collapse/index.html?hpt=hp_t3.

United Nations (2007). *Hyogo Framework for Action 2005–2015: Building the Resilience of Nations and Communities to Disasters*. Geneva: United Nations – International Strategy for Disaster Reduction.

UN. (2012). *Managing Water Under Uncertainty and Risk*. Paris: United Nations.

UN. (2013a). *Global Assessment Report on Disaster Risk Reduction*. New York: United Nations.

UN. (2013b). *Integrating Environmental Sustainability and Disaster Resilience in Building Codes. Annex C: Australia*. Bangkok: United Nations, Economic and Social Commission for Asia and the Pacific.

UN-HABITAT. (2008). *State of the World's Cities 2008/2009*. Nairobi: UN-HABITAT.

UN-HABITAT. (2009). *State of the World's Cities 2010/2011*. Nairobi: UN-HABITAT.

UN-HABITAT. (2013). *State of the World's Cities 2012/13*. New York: Routledge.

UNCED. (1992). *Agenda 21*. Rio de Janerio: United Nations.

UNEP. (2003). *Sustainable Building and Construction: Facts and Figures*. Paris: United Nations Environment Programme.

UNEP. (2006). *Sustainable Building & Construction Initiative*. Paris: United Nations Environment Programme.

UNEP. (2007). *Buildings and Climate Change: Status, Challenges and Opportunities*. Paris: United Nation Environment Programme.

UNEP. (2008a). *Reforming Energy Subsidies*. New York: United Nations Environment Programme.

UNEP. (2008b). *Sustainable Procurement: Buying for a Better World*. Paris: United Nations Environment Programme.

UNEP. (2010). *Common Carbon Metric*. Copenhagen: UNEP SBCI.

Union of Concerned Scientists. (2013). *National Flood Insurance Program Debt Grows*, accessed 7 November 2013 at http://www.ucsusa.org/assets/images/gw/overwhelming-risk-rethinking-flood-insurance/Chart-National-Flood-Insurance-Program-Debt_Full-Size.jpg.

UPI. (2013). *German Energy Minister Hails 'Success' of Solar Subsidy Reforms*, 10 July, accessed 13 November 2013 at http://www.upi.com/Business_News/Energy-Resources/2013/07/10/German-energy-minister-hails-success-of-solar-subsidy-reforms/UPI-52391373428980/.

Urban Green. (2013). *Building Resiliency Taskforce. Full Report*, June. New York: Urban Green Council.

US Department of Energy. (2013a). *Better Buildings Challenge*, accessed 17 December 2013 at http://www4.eere.energy.gov/challenge/home.
US Department of Energy. (2013b). *Better Buildings Challenge. Progress Update*, Spring. Washington, DC: US Department of Energy.
USGBC. (2010). *Green Building Facts*. Washington, DC: US Green Building Council.
USGBC. (2013a). *Incentives and Financing*, accessed 12 December 2013 at http://www.usgbc.org/advocacy/priorities/incentives-financing.
USGBC. (2013b). *Infographic: LEED in the World*, 3 May, accessed 10 December 2013 at http://www.usgbc.org/articles/infographic-leed-world.
USGBC. (2013c). *LEED for Existing Buildings: Operations & Maintenance*. Washington, DC: US Green Building Council.
USGBC. (2013d). *USGBC History*, accessed 12 December 2013 at http://www.usgbc.org/about/history.
Vaidya, C., V. Dhar and N. Dasgupta Sur (2010). 'Knowledge sharing programme under JNNURM'. *Context*, **7** (2), 75–82.
Van Caenegem, R. (2003). *An Historical Introduction to Western Constitutional Law*. Cambridge: Cambridge University Press.
Van der Heijden, J. (2009). *Building Regulatory Enforcement Regimes: Comparative Analysis of Private Sector Involvement in the Enforcement of Public Building Regulations*. Amsterdam: IOS Press.
Van der Heijden, J. (2010a). 'On peanuts and monkeys: private sector involvement in Australian building control'. *Urban Policy and Research*, **28** (2), 195–210.
Van der Heijden, J. (2010b). 'One task, a few approaches, many impacts: private sector involvement in Canadian building control'. *Canadian Public Administration*, **53** (3), 351–74.
Van der Heijden, J. (2010c). 'Smart privatization: lessons from private-sector involvement in Australian and Canadian building regulatory enforcement regimes'. *Journal of Comparative Policy Analysis*, **12** (5), 509–25.
Van der Heijden, J. (2011). 'Friends, enemies or strangers? On relationships between public and private sector service providers in hybrid forms of governance'. *Law and Policy*, **33** (3), 367–90.
Van der Heijden, J. (2012). 'Voluntary environmental governance arrangements'. *Environmental Policies*, **21** (3), 486–509.
Van der Heijden, J. (2013a). 'Interacting state and non-state actors in hybrid settings of public service delivery'. *Administration & Society*, **120** (4), 814–48. doi: 10.1177/0095399713481349.
Van der Heijden, J. (2013b). 'Is new governance the silver bullet? Insights from the Australian buildings sector'. *Urban Policy and Research*, **31** (4), 453-471. doi: 10.1080/08111146.2013.769156.

Van der Heijden, J. (2013c). 'Looking forward and sideways: trajectories of new governance theory'. *Amsterdam Law School Research Paper*, **2013** (4), 1–26.
Van der Heijden, J. (2013d). 'Voluntary environmental governance arrangements in the Australian building sector'. *Australian Journal of Political Science*, **48** (3), 349–65. doi: 10.1080/10361146.2013.821456.
Van der Heijden, J. (2013e). 'Greening the building sector: roles for building surveyors'. *Journal of Building Survey, Appraisal & Valuation*, **2** (1), 24–32.
Van der Heijden, J. (2013f). 'Win-win-win: promises of and limitations to voluntarily greening the building sector'. *Construction Infrastructure Architecture World*, August–September, 80–5.
Van der Heijden, J. (forthcoming, 2014a). 'Experimentation in policy design: insights from the building sector'. *Policy Sciences*, at http://link.springer.com/article/10.1007%2Fs11077-013-9184-z. doi: 10.1007/s11077-013-9184-z.
Van der Heijden, J. (forthcoming, 2014b). 'Regulatory failures, split-incentives, conflicting interests and a vicious circle of blame: the new environmental governance to the rescue?' *Journal of Environmental Planning and Management*, at http://www.tandfonline.com/doi/abs/10.1080/09640568.2014.907135#.U6IurvmSyb. doi: 10.1080/09640568.2014.907135.
Van der Heijden, J. (forthcoming, 2014c). 'What "works" in environmental policy-design? Lessons from experiments in the Australian and Dutch building sectors'. *Journal of Environmental Policy & Planning*, at http://www.tandfonline.com/doi/pdf/10.1080/1523908X.2014.886504. doi: 10.1080/1523908X.2014.886504.
Van der Heijden, J. and J. De Jong (2009). 'Towards a better understanding of building regulation'. *Environment and planning B, Planning and Design*, **36** (6), 1038–52.
Van der Heijden, J. and E. Van Bueren (2013). 'Regulating sustainable construction in Europe: an inquiry into the European Commission's harmonization attempts'. *International Journal of Law in the Built Environment*, **5** (1), 5–20.
Van der Horst, T. and P. Vergragt (2006). 'The seven characteristics of successful sustainable system innovations', in P. Nieuwenhuis, P. Vergragt and P. Wells (eds), *The Business of Sustainable Mobility*. Sheffield: Greenleaf, pp. 125–41.
Van der Heijden, J., H. Visscher and F. Meijer (2007). 'Problems in enforcing Dutch building regulations'. *Structural Survey*, **24** (3/4), 319–29.

VCEC. (2005). *Housing Regulation in Victoria: Building Better Outcomes, Victorian Competition and Efficiency Commission, Final Report*. Melbourne: Victorian Competition and Efficiency Commission.

Verhoest, K., B. Verschuere and G. Bouckaert (2007). 'Pressure, legitimacy, and innovative behavior by public organizations'. *Governance: An International Journal of Policy, Administration and Institutions*, **20** (3), 469–97.

Vermande, H. and J. Van der Heijden (2011). *The Lead Market Initiative and Sustainable Construction: Screening of National Building Regulations*. Bodegraven: PRC and TU Delft.

Vinagre Diaz, J.J., M.R. Wilby and A. Belén Rodríguez González (2013). 'Setting up GHG-based energy efficiency targets in buildings: the ecolabel'. *Energy Policy*, **59** (C), 633–42.

Vine, E. (2005). 'An international survey of the energy service company (ESCO) industry'. *Energy Policy*, **33** (5), 691–704.

Vogel, S.K. (1996). *Freer Markets, More Rules: Regulatory Reform in Advanced Industrial Countries*. Ithaca, NY: Cornell University Press.

VROM. (2003). *Patio Sevilla. Onderzoek naar het instorten van balkons, Ceramique blok 29*. The Hague: Ministerie van VROM and Sdu Uitgevers.

Walker, H. and S. Brammer (2009). 'Sustainable procurement in the United Kingdom public sector'. *Supply Chain Management*, **14** (2), 128–37.

Walker, H. and W. Phillips (2009). 'Sustainable procurement: emerging issues'. *International Journal of Procurement Management*, **2** (1), 41–61.

Walker, N. and G. de Búrca (2007). 'Reconceiving law and new governance'. *Colombia Journal of European Law*, **13** (4), 519–37.

Wall Street Journal. (2013). *Before Dhaka Collapse, Some Firms Fled Risk*, 8 May, accessed 26 February 2013 at http://online.wsj.com/news/articles/SB10001424127887324766604578458802423873488.

Walsh, B. (2013). 'The hard math of flood insurance in a warming world', accessed 7 November 2013 at http://science.time.com/2013/10/01/the-hard-math-of-flood-insurance-in-a-warming-world/.

Weber, M. and J. Hemmelskamp (2005). *Towards Environmental Innovation Systems*. Berlin: Springer.

Webler, T. and S. Tuler (2006). 'Four perspectives on public participation process in environmental assessment and decision making'. *Policy Studies Journal*, **34** (4), 699–722.

WGBC. (2011). *EU Regulatory Frameworks in a Nutshell*. Toronto: World Green Building Council.

WGBC. (2013). *The Business Case for Green Building*. Toronto: World Green Building Council.

Wheeler, S.M. and T. Beatley (2009). *The Sustainable Urban Development Reader*, 2nd edn. London: Routledge.
Whelan, R. (2012). Submission: Recent trends in and preparedness for extreme weather events. Letter to the Senate Standing Committees on Environment and Communications. Insurance Council of Australia, Sydney.
Wiklund, H. (2005). 'In search of arenas for democratic deliberation: a Habermasian review of environmental assessment'. *Impact Assessment and Project Appraisal*, **23** (4), 281–92.
Wilkinson, M. (2010). 'Three concepts of law'. *Wisconsin Law Review*, **2010** (2), 637–718.
Williamson, O.E. (1996). *The Mechanisms of Governance*. New York: Oxford University Press.
Wilson, J.Q. (1989). *Bureaucracy: What Government Agencies Do and Why they Do It*. New York: Basic Books.
Witztum, A. (2005). *Economics: An Analytical Introduction*. Oxford: Oxford University Press.
Wong, R. (2011). 'Solar potential of HDB blocks in Singapore'. *ESI Bulletin*, **4** (3), 6–7.
World Bank. (2008). *Building Resilient Communities*. Washington, DC: World Bank.
World Bank. (2009). *Climate Resilient Cities: A Primer on Reducing Vulnerabilities to Disasters*. New York: World Bank.
World Water Council. (2007). *The Struggle for Water*. Marseille: World Water Council.
Wright, M. (2013). 'Paving the way for subsidy-free solar', 14 October, accessed 13 November 2013 at http://www.businessspectator.com.au/article/2013/10/14/solar-energy/paving-way-subsidy-free-solar.
Wurzel, R., A. Zito and A. Jordan (2013). *Environmental Governance in Europe*. Cheltenham, UK and Northampon, MA, USA: Edward Elgar.
Wynn, G. (2013). 'Falling solar costs allow a new approach to subsidies', 16 October, accessed 13 November 2013 at http://www.reuters.com/article/2013/10/16/column-wynn-solar-cost-idUSL6N0I51VP20131016.
Yates, A. and A. Bergin (2009). *Hardening Australia: Climate Change and National Disaster Resilience. Special Report 29*. Canberra: Australian Strategic Policy Institute.
Yates, D. (2003). *Weathertightness of Buildings in New Zealand. Report of the Government Administration Committee's Inquiry into the Weathertightness of Buildings in New Zealand*, March. Wellington: Government Printer.
Young, O., L. King and H. Schroeder (eds) (2008). *Institutions and Environmental Change*. Cambridge, MA: MIT Press.

Yu, S., M. Evans, P. Kumar, L. Van Wie and V. Bhatt (2013). *Using Third-party Inspectors in Building Energy Codes Enforcement in India.* Springfield, VA: US Department of Commerce.

Yu, Z., W.L. Zhang and T.Y. Fang (2013). 'Impact of building orientation and window-wall ratio on the office building energy consumption'. *Applied Mechanics and Materials*, **409–410**, 606–11.

Yudelson, J. and U. Meyer (2013). *The World's Greenest Buildings.* Abingdon: Routledge.

Zarsky, L. (1997). *Stuck in the Mud? Nation-states, Globalization and the Environment.* The Hague: Nautilus Institute.

Zeitlin, I. (1997). *Rulers and Ruled: An Introduction to Classical Political Theory from Plato to the Federalists.* Toronto: University of Toronto Press.

Zhang, S., J. Teizer, J.-K. Lee, C. Eastman and M. Venugopal (2013). 'Building Information Modeling (BIM) and safety: automatic safety checking of construction models and schedules'. *Automation in Construction*, **29** (1), 183–95.

Zimring, M., M. Borgeson, A. Todd and C. Goldman (2013). *Getting the Biggest Bang for the Buck.* Berkeley, CA: Lawrence Berkeley National Laboratory.

Zografakis, N., A. Menegaki and K. Tsagarakis (2008). 'Effective education for energy efficiency'. *Energy Policy*, **36** (8), 3226–32.

Index